Joseph Agassi

Radiation Theory and the Quantum Revolution

Birkhäuser Verlag
Basel · Boston · Berlin

Author:

Professor Joseph Agassi

Department of Philosophy
Tel-Aviv University
Tel-Aviv 69978
Israel

Department of Philosophy
York University
4700 Keele Street
North York, Ontario
Canada M3J 1P3

A CIP catalogue record for this book is available from the Library of Congress, Washington D.C., USA

Deutsche Bibliothek Cataloging-in-Publication Data

Aggāsî, Yôsēf:
Radiation theory and the quantum revolution / Joseph Agassi.–
Basel; Boston; Berlin: Birkhäuser, 1993

This work is subject to copyright. All rights are reserved, whether the whole or part of the material is concerned, specifically the rights of translation, reprinting, re-use of illustrations, recitation, broadcasting, reproduction on microfilms or in other ways, and storage in data banks. For any kind of use, permission of the copyright owner must be obtained.

©1993 Birkhäuser Verlag, P.O.Box 133, CH-4010 Basel, Switzerland
Softcover reprint of the hardcover 1st edition 1993
Camera-ready copy prepared by the author
Printed on acid-free paper produced from chlorine-free pulp
Cover design: Markus Etterich, Basel

ISBN-13: 978-3-0348-7217-1   e-ISBN-13: 978-3-0348-7216-4
DOI: 10.1007/ 978-3-0348-7216-4

9 8 7 6 5 4 3 2 1

to my children

CONTENTS

Preface . . . . **XI**

Acknowledgement . . . . **XVII**

Chapter 1:   **Radiation Theory** 1

    Absorption And Emission 1
    Big Questions And Small Questions 3
    The Rear Guard of Science 5
    The Background to Radiation Theory 8
    Unity And Diversity In Nature 12
    Kirchhoff's Law 15

Chapter 2:   **The Background to Radiation Theory** 16

    Flames as Things 16
    Heat as Substance 18
    Radiant Heat 20
    The Place of Prevost's Law in History 25

Chapter 3:   **The Rise of Spectroscopy** 29

    Spectral Lines 29
    The Discovery of Spectral Lines: A Problem 31
    The Discovery of Spectral Lines: The Story 35
    The Discovery of Spectral Lines: A Discussion 38
    The Rise of Astrophysics 41
    Clues and Promises in Science 46
    The Place of Young in History 49

Chapter 4: The Changing Scenery **52**

    The Wave Theory of Light 52
    More About Waves 57
    Light Waves and Matter 60
    Heat as Energy 63

Chapter 5: Kirchhoff's Law **67**

    Spectral Analysis 67
    Absorption Spectra 71
    Emission and Absorption Coefficients 75
    Kirchhoff's Law 80
    Kirchhoff's Followers 85
    Spectral Lines Between Kirchhoff and Bohr 88
    Atomic Spectra and the End of Atomism 91

Chapter 6: The Background To Quantum Theory **93**

    The Stefan-Boltzmann Law 93
    Wien's Law 96
    The Red Herring of the Violet Catastrophe 100
    Planck and Bohr on Models 102
    Planck's Law 108
    Einstein and the Photoelectric Effect 112
    The Crisis in Physics 115

Appendix A: The Kirchhoff-Planck Radiation Law 117

    Prevost's Law of Exchange 118
    Fraunhofer's Discovery of Spectroscopy 119
    Stewart's Law of Radiation 120
    Preliminaries to Kirchhoff's Law of Radiation 121
    Kirchhoff's Law and Its Proof 123
    Between Kirchhoff and Planck 125
    Planck's Studies Prior to His Quantization 126
    Einstein's Version of Kirchhoff's Law 129
    References and Notes 132

Appendix B: The Structure of the Quantum Revolution 139

    Kuhn on Planck 139
    Kuhn on the Quantum Revolution 140
    Kuhn's Sociology of Science 142
    Planck's Program 144
    The Status of Entropy 146
    Planck Versus Boltzmann 148
    Planck's Capitulation to Boltzmann 152
    Conclusion 154
    Notes 155

Appendix C: Quantum Duality 156

    Introduction and Abstract 156
    Dismissing Popular Cynicism about Science 156
    Approximationism in Action in Modern Physics. 160
    The Confused Roots of Complementarity 164
    Conclusion 168

Bibliographic Note 169

Name Index 171

Subject Index 175

PREFACE

This book tells the story of the researches that are traditionally lumped together under the label "radiation theory" and revolving, loosely speaking around the familiar heat-and-light exchange (hot bodies emit light or radiate; the absorption of light, especially sunlight, is warming). This characterization, we will soon find out, is too crude. Attempts to improve upon it have brought about the revolution in physics known as the quantum revolution (because the revolutionary change involved was the chopping up of light waves into discrete quantities) early in the twentieth-century. I wrote it with the following rules in mind:

(a) try to present the development of the intellectual background relevant to your story;

(b) try to present the details of your story in a critical manner; try to present developments as results of dissatisfaction with current states of affairs;

(c) try to avoid reporting any piece of information without explaining what purpose it serves.

The first rule originated by Meyerson, Burtt and Koyré; the second originated by Popper and Koyré (Koyré refers to Bachelard in this context); the third is plain common sense though seldom practised (Koyré is the model here); being common sense it is understandably almost nowhere stated; the regular infringement on it made me state it and discuss it in some detail in my first monograph, *Towards an Historiography of Science, History and Theory, Beiheft 2*, 1963. Now the rule,

(d) contrast approaches and opinions, unorthodox and orthodox ones, heresies and received opinions,

is a corollary to the above three rules. I do not know if it is valid. Shortly before his death Koyré kindly encouraged me, suggesting that criticizing contemporary historians of science should be avoided as it sounds hostile and its targets are often better ignored. This comment is common and regrettably true. Regrettably criticism and disputations do sound hostile; I wish I could make it sound friendly, but I do not know how. Criticism and contrasts of opinions should not sound hostile, even when regrettably launched in a hostile manner and out of hostility, since criticism and contrast are what make reading more easily comprehensible. But certainly it should be possible to launch friendly criticism without the usual sugar-coating of insincere and distracting flattery.

To that end conventions should be altered; convention should require that critics praise the opinions they criticize whenever possible and only sincerely, not as sugar coating; convention should require that criticism be taken as a form of tribute anyway, even if it be a tribute to an opinion as the received opinion. This can easily be implemented by organizations recognized as setting standards, such as the Nobel Committee or the Royal Society of London. As it happens, that Committee emulates that Society; it was the pioneering act of that Society, in the 1660's, to institute the current conventions, and this is why the founding of that Society marks the beginning of the scientific era. Yet the standards were defective, and, indeed, they were periodically reformed. One of the Society's unintended measures was to judge all contrasts between opinions and criticisms of opinions as hostile acts; this was unintended, and it was a consequence of the Society's ruling that conjectures should be avoided as much as possible, so as to prevent the hostilities subsequent upon the criticism pitched against them. A corollary to this was that philosophical speculations should be avoided altogether. As that Society has broken from its rules and elected as a member a philosopher as a philosopher, and as that philosopher, Sir Karl Popper, is of the Socratic opinion that all criticism is a form of paying respect, there is still hope.

Please permit me one added paragraph of a personal *apologia*. Like all who fervently wish to see a reform, I suppose I regrettably sound somewhat priggish: why not let things be? why stress the unfairness of this or that historian to this or that historical figure? I find the reform of the etiquette of scientific research and the reform of the method of writing the history of scientific research important and interwoven: defenders of the current views on scientific research and on the etiquette that goes with it see a tremendous support for these views in the literature on the history of science, especially the popular stories which it incorporates and which ring so true. This is why I wrote my already mentioned first monograph, where I illustrate repeatedly that the moral from any study in the history of science is the moral usually brought to it by its author, that most authors in the field consider their task to be the revalidation of science as they comprehend it, that consequently they are terribly uncritical and are ready to be unfair to any past thinker who violate their own ideas on research, either by censuring them or by condescendingly overlooking their supposed transgressions. Thus, the revalidation of the current views on science regularly found in it is not any profound lesson learned from history but the submission to cheap propaganda. While rewriting this book repeatedly, I derived much assistance from individuals who are friends, esteemed colleagues and learned critical thinkers (some of them are mentioned in the *Acknowledgement* below). Most of their comments were corrections of errors, confusions and wrong presentations, of course; other of their comments, no less welcome, drew my attention yet again to the need to take cognizance, while writing a history of science, of so much cheap propaganda that causes so much misunderstanding of relatively straightforward texts. My forays into the history of physics are not motivated by a wish to explode some tributes to some great thinkers of the past as myths, nor to restore fairness to other great thinkers of the past whose works are unjustly overlooked or censured; presumably they cared more for science then for what future historians might say about them. My concern in rectifying some historical details lies elsewhere: I find the current ideas about scientific research intimidating and boosted by wrong histories. Perhaps this is my excuse for having failed in my feeble efforts to participate actively in current scientific research. Perhaps this is why I repeatedly single out the popular historian and philosopher of science, Thomas S. Kuhn, as a target for my aspersions: we both find intimidation as an integral part of contemporary scientific training, yet he openly defends it (see his "Replies to My Critics" in the best-seller *Criticism and the Growth of Knowledge*, edited by Imre Lakatos and Alan Musgrave, Cambridge University Press, 1970). As a physics student I could not understand much of what my professors were teaching me, yet it was the material that was presented to me as background to quantum theory that was the barrier that prevented me from doing well. This led me years later to my study of the Kirchhoff-Planck radiation law. I wrote this book in an effort to understand and explain the situation in an effort to reduce the intimidation in future training of physicists. So let me speak of intimidation in general.

Most of the material in the history of science is meant for popular consumption. Most of works in this literature intimidate their intended readers. (Also much of the expert literature is, in physics and in its history.) Intimidation is unacceptable in principle; in the case at hand it is also unacceptable as it makes popular and/or introductory science something inferior, whereas in truth science traditionally boasts exotericism, openness to every interested member of the general public, so that nothing is truly scientific until it becomes accessible to them with relative ease.

    This book is an effort to close a gap, then, to present a segment of the history of science in a manner that accords with the view of it as exoteric, as open to the general public, and as a critical activity, where thinkers regularly criticize their predecessors so as to improve upon their ideas. This book also addresses the knowledgeable, including scientists and historians of science, yet it mainly addresses science students (since my research and writing were executed with an eye on the difficulties which I had met over forty years ago as a science student, and which impeded my progress).

    I cannot boast much success in the execution of this book: the last chapter is full of gaps due to my decision not to require of the reader any mathematical knowledge. Readers concerned with these gaps may easily find more information in the literature, including my "The Kirchhoff-Planck Radiation Law" (*Science*, 156, 1967, 61-67), reproduced here as Appendix A. The last two decades have seen a development in the field of the history of science. Some young authors write while contrasting different answers of different historical figures to the same question, indicating improvement and thus a growth of knowledge. Most of them also appeal to the authority of Thomas S. Kuhn. This is strange since Kuhn opposes such contrasts. In his view there is no reasonable contrast between different answers to the same question, between alternative scientific theories which belong to different periods. This is his celebrated doctrine of "incommensurability", expounded in his best-seller *The Structure of Scientific Revolutions* (1962) and elsewhere. Admittedly, Kuhn has softened his stance when he said that some contrasts can be reasonably made, some not; this is hardly significant, unless it is supplemented with a criterion as to what contrasts are to be judged reasonable and what not. What is his criterion of reasonable contrasts? The discussion of this question fills volumes. For my part I can say I still do not know Kuhn's criterion despite some efforts to comprehend him, and I think he himself does not care about it overmuch, that the kernel of his message is the revalidation of the socially given, of the community of science as it is. The interested readers are referred to my *Science and Society* 1981, where it is discussed at length and shown rather frivolous.

There is much primary and secondary material on the developments described here, and it is mentioned in the *Bibliographical Note* in the end of this volume. For a general background nothing beats the first volume of Sir Edmund Whittaker's *History of Theories of the Aether and Electricity*, despite its author's objectionable philosophy which has repeatedly led him to offer distorted readings of historical material. William McGucken has offered an original, detailed, scholarly study of much of the material leading to the quantum revolution in his *Nineteenth Century Spectroscopy, 1802-1897*, 1969. The very last steps leading to the revolution and on through it, were studied by Martin Klein and by Thomas S. Kuhn; they are blow-by-blow, very scholarly descriptions. My disagreements with them in historical detail are discussed in my extensive review of Kuhn's book ("The Structure of the Quantum Revolution", *Philosophy of The Social Sciences*, 13, 1983, 367-81), reproduced here as Appendix B. There is no explicit statement of Klein's views on the story in general. As to Kuhn view, I have discussed it in that review. His paper which is devoted to a detailed response to his critics ("Revisiting Planck", *Historical Studies in the Physical Sciences*, 14, 1984, 231-52, reprinted in Stephen Brush, editor, *History of Physics: Selected Reprints*, College Park MD, American Association of Physics Teachers) dodges that review and the criticism it contains. An established rule of the commonwealth of learning exempts me from further discussing his work.

My hope is that the present book will be read without effort and contribute to the reader's comprehension of the background to quantum theory in a manner that would have helped me had I been exposed to it as a physics student, and to the reader's readiness to develop a critical attitude to science without fear of seeming unfriendly to science: science is the biggest show on earth, and it needs no defense.

The present book discusses, first and foremost, questions concerning the history of science. Some of them are traditional, such as, how did Newton miss discovering the fact that parts of the spectrum are dark (the spectral lines, as they are called), how did Young come to revive the wave theory of light and what was the obstacle he could not overcome, and how did the materialist theorists of heat ignore the argument that heat is created by friction, and why was the use of spectral lines for spectral analysis delayed by nearly half-a-century? Other are new, such as, why were the followers of Kirchhoff apologetic about their investigations of his law, and why were they thrilled by it, and what has caused the crisis in physics?

I repeatedly discuss traditional views on any concern I raise and try to improve upon them. Kirchhoff's reasoning, which was initially as tortuous as often new thoughts are, and which is grossly misrepresented in texts of both physics and its history, is here presented in a manner that enables readers to follow him with ease. As my chief concern is to overcome the very difficulties to understand the matter (which were not removed by reading of the primary and the secondary literature on the topic), I hope that readers will appreciate my presentation when they see what was so intriguing in the thinking that went into the research into spectroscopy and what made it so exciting. They are invited, of course, to criticize and improve upon my presentation, and thus leave behind the habit of forcing down the throats of uncomprehending students this (or any other) episode in the history of physics.

In line with this general aim I have added a final appendix, Appendix C, on the most puzzling aspect of quantum mechanics, the quantum wave-particle duality. I am afraid Appendix C assumes some familiarity with the theory. Yet even other readers may see that that Appendix C employs the same technique which is used throughout this book, and which enables me to do so much better than writers more qualified than I am as physicists or as historians.

The reason I can do so much better than they is that while I use their insights freely, I avoid one activity that engages them and takes too much effort and obscures the rationality present in the historical material: I do not try to defend the great thinkers of the past against the accusation that they were in error, since, put very simply, *error is not culpable and needs no defense* especially if it is useful but even if it is not. Also, interesting errors may prove to be only interesting: they are often conducive to the growth of knowledge. Truth, said Bacon, emerges quicker out of error than out of confusion. What makes science so exciting is that its progress is not linear, that it is an ongoing disputes between different opinions which are repeatedly improved while one party borrows as much as possible from the better ideas of its opposition and while always accepting criticism and attempting to get free of older errors. But there is no error-free science. Science is a set of great ideas that need not be defended on the pretext that it is error-free. For more details see my *Towards an Historiography of Science, History and Theory, Beiheft 2*, 1963, facsimile reprint 1967.

Herzliah, Israel, Passover, April, 1990;
North York, Canada, Jewish New Year, September, 1990.

ACKNOWLEDGEMENT

Some thirty years ago I read a paper on Kirchhoff's law which was soon published—after Abner Shimony and Martin Kline had kindly offered helpful comments. It is here reissued as Appendix A. L. Pearce Williams of Cornell University heard the presentation and kindly invited me to write this book. He offered me a contract for the history of science series which he then edited.

The series' publisher folded and the book failed regularly to find an alternative home: most of the publishers I approached returned the manuscript with the usual bland excuses; others were honest and open enough to report dismay at my expression of disrespect for professors who are stuffed-shirts (see Chapter 1 below); I sincerely regret I could not concede, since this hindered publication. I have with me a publisher's referee's report on an earlier version of this book, full of praise and recommending that the book remain unpublished all the same because it is so unorthodox in so many ways. I gave up hope.

Michael Segre of the Institute for the History of Science of Munich University came to the rescue. He showed initiative, submitted this work to the publisher, read the manuscript and made extensive useful comments that caused much rewriting and consultations. Samuel Goldsmith of Tel-Aviv University and Mendel Sachs of SUNY Buffalo then showed interest, corrected my faulty physics, and offered wise advice. Zev Bechler and Menachem Fisch of Tel-Aviv University read the final version and made many suggestions. Lev Vaidman of Tel-Aviv University read the essay on quantum duality here presented as Appendix C. Lance Odland of York University corrected the proofs and eliminated the rough edges of the penultimate draft. The publisher's editor, Benno Zimmermann, promised to read his referees' report for informative responses, and ignore any expression of hostility—which any of them might include. This he did.

Appendix A was previously published in *Science*, 156, 1967, pp. 61-7. Appendix B was previously published in *Philosophy of the Social Sciences*, 13, 1983, pp. 367-381. Both appear here with the kind permission of the respective editors.

My gratitude to all of them.

Chapter 1

RADIATION THEORY

1.1   Absorption And Emission

The name "Radiation Theory" denotes those researches, theories, experiments, and problems—chiefly problems—which occupied physicists and chemists and astronomers between around 1800 and precisely 1900 and which gave way to quantum theory at the end of 1900 or at the latest in 1913 when quantum theory was established. The term "radiation" denotes in English both the process of radiating, the process of emitting light, and the substance radiated, light itself. Here we will be concerned chiefly with the process, though, of course, there is a strong interaction between the two questions, what is the process of radiation? and, what is the light that is radiated? Further, the terms "emit" and "radiate" will be used here as synonyms, yet the modern term "radiation" is applied to the earlier researches of emission and absorption quite insensitively; it is considered common knowledge today that radiation conveys energy with or without matter, and that pure energy is electromagnetic radiation which is interpreted as light. This was far from the minds of the early researchers and their struggles with radiation has helped develop the modern ideas of energy.

Some sources of light are the sun and stars, flames, hot metals, and even fireflies. Mirrors are not sources of light; we prefer to view them as reflectors, not as genuine sources. This makes every allegation as to a source of light problematic: is the moon a source of light or a reflector of the sunlight? It turns out to be a reflector, not a source. Ancient astronomers said so as they knew that the shining side of the moon always faces the sun. Radiation theory reveals more: moonlight is the same kind of light as sunlight reflected from a stone mirror.

Eyes of cats and dogs shine in the dark; are these eyes sources or reflectors? This question illustrates the difficulty of deciding when light is emitted from a source and when not. It is not only the lack of information that hinders decision; since light is a wave phenomenon of sorts, there is such a thing as light resonance, akin to sound resonance, and it is something resembling both reflection and emission. Echo is sound reflection, but when a sound instrument, such as the concert-piano's wing, resonates, it absorbs sound and then emits sound of the same pitch; this also happens in the case of light; should resonance be considered emission or reflection? Why? and how do we decide when a body reflects and when it resonates?

Perhaps surprisingly, metal radiates not only when hot:

every object radiates at every possible temperature.

The quantities of radiation emitted by objects at room temperature and below have been carefully measured. Of course, the detection of light coming from the ordinary objects at room temperature is executed with the aid of special instruments. That this radiation is not visible proves that what we see and what there is as a matter of fact need not be the same.

So much for the sources of light. Consider now the light beams themselves; let us envisage them as missiles, say bullets. A bullet may pass through an obstacle such as flesh and get stuck in an obstacle such as bone, or be deflected or ricochetted from an obstacle such as a rock. It looks as if there are no other possibilities, and for the time being we shall ignore them even though they do exist. When a body is transparent, light passes through it. Opaque bodies usually absorb some rays and reflect others; for example, grass absorbs red light and reflects green light. This is not only a roundabout way of saying that grass is green, it is also to say that in red light grass looks black. In the color pictures composed of triple slides, each lit with a different color, the grass looks black on the picture lit with red light and white on the one lit by green light. Similarly, in the violet light of mercury street lights red lips look eerily black.

The optical information on each body, then, should include at least the details about what colors it is transparent to, what colors it absorbs, reflects, emits. Each object is partly transparent, partly opaque, partly absorbent, partly reflecting, and also partly radiant. There are no bodies that are utterly transparent, opaque, or reflecting; for each body there are degrees of these qualities, and they differ for different colors.

Let us return to our missiles. It is not true that flesh is transparent to bullets: much flesh will stop a bullet, especially a slow one; indeed, the flesh slows down a bullet. Each millimeter of flesh a bullet travels through (or traverses) reduces its speed by ever so little, and after sufficient length is traversed, the bullet is slowed down to a complete halt. This is exactly what happens to bullets going through bones, or sandbags, or concrete walls—only the slowing down (the deceleration) is greater in the case of bones and still greater in the case of sandbags. But even water slows down bullets. Now consider light travelling through such a transparent body as a glass of water. Water is not entirely transparent; the darkness of pictures taken at the bottom of the sea in broad daylight clearly illustrates this fact. We may conclude, then, that water slows down light-rays just as it slows down bullets.

This is a nice theory. Experimental examinations showed it to be false. A faint glimmer of light which succeeds to find its way to the bottom of the sea travels along its path as swiftly as it did soon after it had penetrated the surface.

What is the way of thinning a barrage of missiles without slowing them down? Consider a barrage of arrows zooming through a thick forest. An arrow either travels unimpeded, or a tree-trunk stops it dead. The chance of an arrow grazing a tree-trunk is so slim that we shall pretend it is nil. All bodies are transparent to light like the forest to arrows; only some of them are thick forests, some thin. To be more precise, the same body, say a piece of blue glass, is transparent to blue light like a thin forest but to red light like a very thick forest. Viewing matter as a collection of atoms makes it easy to view each of them as a possible target of a light-missile, instead of a tree-trunk. Let us liken the atom to a warrior: it may emit light, it may absorb light, it may reflect light, it may let it zoom by; apparently there are no other possibilities. (There are: some light is deflected. Also, at times a soldier acts as a resonator: when shot at, he shoots back. More often he starts shooting because arrows are flying around; this is induced emission, not resonance.) So, we assume the existence of emission-probabilities and also of absorption-probabilities; the higher the absorption-probabilities of given matter towards a given color, the more likely it is to act like a thick forest towards it.

And so, atoms emit light, absorb it, reflect it, or let it zoom by. Why? How? Under what conditions? These questions are shared by the older radiation theory and the younger quantum theory. As they share questions, the younger replaces the older. Also, the younger answer meets more questions than the older. Example: why does hot metal radiate? and why is grass green? Radiation theory succeeds in answering only one; quantum theory answers both.

1.2 Big Questions And Small Questions

Why is grass green?

This question will not be answered here since this is too difficult. Let us, rather, reflect on the kind of question it is. It is, obviously, the kind of question which young children tend to ask but merely because children ask questions indiscriminately. Adults suppress children's questions as some questions must be suppressed, at least tentatively, because life is too short to allow us to investigate all questions which cross our minds. But some questions should invite attention. In other words, only the suppression of questioning as such should be condemned, not the suppression of some questions in favor of other ones: questioning must be selective.

How should questions be selected? What is the rule or criterion of selection? This is a question on which philosophical schools are violently divided. Let me begin with the answer to it which was once popular amongst the learned but is now popular only amongst the superstitious.

In the Middle Ages it was common to wonder—to ask questions—about the wondrous, about the unusual, the way-out. It was deemed silly to ask, for instance, why are calves usually born with single heads? It was considered much more natural to ask, how come sometimes, however seldom, calves are born with two heads each? The unusual, the so-called monster of nature, was the thrilling, the interesting, the clue to the secrets of nature. The scientific revolution has done away with this.

The superstitious consider it scandalous that scientific researchers systematically ignore extremely unusual and puzzling clues to the secrets of nature.

There are photographs of unusual events, reports of people with special powers, of people who rose alive from their graves; surely, the superstitious argue, here are thrilling cases for science to investigate with hopes of rich findings. But science, following a tradition over three centuries old, will not touch such items.

Phenomena or events which scientists investigate must be repeatable. This is a rule proposed by Galileo Galilei and laid down by Robert Boyle. It is one of the severest rules of the scientific tradition. Ever since the rise of the scientific tradition there were reports of observations that were not found repeatable; science is simply bound to ignore them. Within science unrepeatable reports are not considered true, nor are they considered false; within science they are simply ignored—the rule within science is that we should suspend judgment about them.

This, however, does not help much; on the contrary, there are fewer unusual phenomena than usual ones, of course, and so the elimination of questions concerning the unusual still leaves myriads, such as, why is grass green?

One might suggest a compromise answer. Though the phenomena studied scientifically must be reproducible, some phenomena may still be less usual than others and thus deserve more attention. For example, the appearance of comets is rarer than the appearance of planets; eclipses of the sun or the moon are obviously more unusual than a sunrise; the flooding of the Nile in the spring is more unusual than the flooding of rivers during the rainy season. Fireflies are fairly unusual—more puzzling, perhaps, than a burning, live, red hot piece of coal or a black piece of coal or green grass.

This compromise solution does not work. On the contrary, if the history of science is any indication, the indication goes the other way. True, the riddle of the spring-flooding of the Nile can be tackled fairly directly. But comets and eclipses could only be explained after satisfactory answers to the commoner questions concerning the orbits of the planets were found. The light of fireflies could be explained only after the emission of light by live coal was successfully explained.

A scientific researcher is one who has the capacity to wonder at the very obvious, it seems! But then, so much can be wondered about! How do we select what to invest efforts of research in? What should one study?

There are two modern philosophical schools of thought here. Crudely, the one says, study big questions first, the other says, study small questions first. The big questions school and the small question school each defends its position on technical grounds. René Descartes, the great French thinker of the second quarter of the seventeenth century and the father of modern mathematics, was in favor of tackling the big questions first, the metaphysical questions so-called, such as, what are things made of? He said, once we have answered the big questions, we will be guided by our solution and know which small questions to ask and how to try to answer them. It is very important to notice that Descartes was not opposed to small questions; on the contrary, he proposed that every question should be broken down to its smallest component; but he said that only after answering some general question do we have a reasonable framework in which to tackle small questions. (His proposal to break down everything to its smallest component is very important and very famous; it is known as the mechanistic philosophy.) His predecessor, Sir Francis Bacon, the great English thinker of the first quarter of the seventeenth century, whose inspiration sooner or later led to the foundation of the Royal Society of London in 1660 and to the rise of technological society in the 19th and 20th centuries, defended the view that we must start with small questions. He was aware of the existence of intellectual frameworks and he conceded that they can serve research, but he considered their allure dangerous. We can never be sure, he said, that we answer the big questions correctly without a lot of preliminary work. And if we answer the big questions incorrectly we may easily take the wrong path and, instead of developing science, find ourselves in a quagmire. The preliminary work necessary before answering the big question concerns, of course, small questions which we can safely handle and to which we can find the answers with the aid of humble experiments. It is very important to notice that Bacon was not opposed to big questions; on the contrary, he proposed that every question should be reached in due course, after the smaller questions had been successfully answered; research should begin with the smallest questions and not cease until the largest questions are reached and successfully answered. (His proposal to generalize in smallest steps is very important and very famous; he called this "the ladder of axioms".) The culmination of research, he said, should answer the greatest question and capture scientific metaphysics.

Bacon's influence was tremendous and most beneficial, but these days it is an excuse for an astonishing dismissal of questions. It is regrettably not uncommon today to find scientists, even important scientific researchers, who lose sight of questions and perform very expensive experiments more or less because they fit a vaguely recognized norm. This is so common that these scientists are called normal and their science is called normal science and it is alleged that they pick some normal tasks to solve with the aid of received scientific theory in their fairly routine experiments, and these are called "puzzles". The description and terminology were introduced by Thomas S. Kuhn. He greatly approves of the procedure, which he describes; he calls it "normal", to suggest that it is normal. It is not; the top philosopher of science Sir Karl Popper says flatly, normal science is so submissive it is hardly science at all.

1.3 The Rear Guard of Science

It was Sir Francis Bacon, who urged people to observe such simple facts as mentioned here—the sky is blue, grass is green, gas light is blue. Without Bacon's stress on humble research it is doubtful that people would have volunteered to contribute experiments to the advancement of science; without Bacon's inspiration, without his promise that the large edifice of science will be built out of small building blocks, it is doubtful that we would have today science as we know it. Yet history shows that too many small contributions were left unused in the construction of the edifice of science. The importance of a piece of scientific research seldom increases in time; usually the importance of even a minute discovery is either noticed within a relatively short time or else it gets forgotten. Even if it is unjustly underrated, this is seldom corrected in later periods, and the discovery has to be made again later on. (The case of the discovery of spectral lines is a very striking example; it will be related later on as they constitute a central item in radiation theory.) Usually, however, a piece of research is more often overrated than underrated and its market value, so-to-speak, slowly peters out. For, some stuffed-shirt professors of science who happen to occupy prominent positions in the scientific world and who are in the habit of backing one item of research or another. If this turns to be an error—and we all err from time to time—then they prefer to have their error slowly sink into oblivion rather than stand corrected. Conversely, the history of science is sometimes, though by no means always, so very dramatic just because a newly discovered fact or idea which these same stuffed-shirt professors of science who occupy prominent positions in the scientific world underrate yet it grows in significance despite all semiofficial opposition. Later on the official history of the discovery is written up and it corrects the past and removes the human drama. But even these dramatic cases are in no way instances such as those which Sir Francis Bacon had in mind when he advocated humble contributions to science.

Bacon and Descartes hardly imagined that the world of science could be occupied, let alone dominated, by stuffed-shirt professors of science who occupy prominent positions in the scientific world. The professors in the times of Bacon and Descartes were old-fashioned clerics who had nothing at all to do with what these two thinkers would consider science. They both took it for granted that all those who care for science are always appreciative of whatever is (obviously) valuable in science and never expected the old-fashioned cleric-professors to begin to appreciate science. When Bacon said that an item of science may increase in value, he meant that its objective value would increase, not the appreciation accorded to it by researchers. He clearly did not have in mind stuffed-shirt professors who occupy prominent positions, as they would not listen to the young researcher with burning eyes. It did not occur to Bacon that some young bright researchers would have to fight for fair hearings and would not be heard until they became middle-aged, wrinkled, and somewhat embittered. But at times such are the sad facts. The last student of radiation-theory and the father of quantum theory, is a point in case: Max Planck wrote his scientific autobiography after he had won the highest recognition anyone at the time could have gained for scientific research, yet the autobiography is bitter. Even Planck's (justly) admired teachers, whose work he continued all his life, were stuffed-shirt professors of science who occupied prominent positions in the scientific world. They are all, all stuffed-shirts, Planck bitterly suggested. When I retell this rather well-known story to my colleagues, they declare all this past history. Today, they tell me, there is no problem about what questions to solve, small or big, and so there is no room for stuffed-shirts to adjudicate; today, I am reassured, the problem-situation is radically altered; research has grown large; most individual researchers belong to large teams; though individuals tackle small problems, I am told, their large teams tackle large problems. Answers to small problems accumulate this way, I am assured, and are turned by teams into answers to the really big ones.

CHAPTER 1

Dear reader with burning eyes, do not believe this story. But do not dismiss it either. Never underestimate the strength of the self-satisfied occupants of prominent positions, not even in the scientific world! Their strength lies in their good reason; they do have a good reason for their self-satisfaction: they are the guardians of science, which is the best and the most prestigious enterprise.

Some historians of science love to poke fun at stuffed-shirt professors of science who occupied prominent positions in the safely distant past, signalling that we are now free of this ill. They pretend that now science is in the best shape possible. They thus foster stagnation.

Things are not in good shape from the very start; we ask big questions and small questions; almost like children, we ask indiscriminately; we answer all sorts of questions, and almost always too hastily. In the history of science almost all reports follow the same pattern: experiments were performed less carefully than required, researches were not satisfactorily completed and books were published without expressing their authors' ideas well enough. Life is short; in haste we ignore much that we would rather attend to more carefully.

What is to be done? It is not enough to attend to a small question or to a big question as we may get stuck with our choice, and the choice may be wrong. Checking to see if different answers to different questions are in sufficient accord with each other may be helpful.

As it happens, this is where radiation theory broke down, in the fitting together of different parts of physics: heat and light convert to each other, electricity and light do too, and so do heat and electricity; these three stories should accord with each other, and they did not. Such breakdowns are almost inevitable. Some thinkers try to match different answers but, as ever, they do so in haste; some answers to big questions look as if they match well enough with some answers to small questions. Success, even seeming success, brings about understandable self-congratulation. Here, stuffed-shirt professors declare solemnly, here is a big answer well-prepared, patiently evolved from well-prepared experiments; here answers to the small questions have led to the answer to the big questions. Self-satisfied professors of science defend the status-quo. Fortunately, there are different kinds of people: some of them, professors or not, researchers or not, teachers or not, famous or not, may admire things as they are, yet be less easily pleased with them. Fortunately, most of those concerned with science are not the stuffed shirts that the guardians of science are. These guardians are admittedly harmful to some extent, as the dogmatic defense of a good idea may be as pernicious than the dogmatic defense of a poor idea. Also, the dogmatic defense of science makes the study of its history quite impossible as it makes historians write while taking care not to displease professors of science. Answers to big questions lead to small questions, and answers to small questions are likely to upset received answers to the big questions. And so the big professors must be on guard—they think—against the wild young inquirer with burning eyes. In our case, the big questions concern the nature of matter, of heat, of light. While radiation theory investigated small questions, concerning red-hot metals and yellow flames, the answers to the big questions kept changing. Imagine the answers to the big questions to be the scenery on a stage, the small ones as bit actors. Some historians have no patience with small actors, as the large stage scenery fully captures their fancy. Other historians, on the contrary, declare this cheap and lazy: they choose small details and study them with devotion. The dissatisfaction with both kinds of history of science springs from viewing it as a play in which the actors arrange and rearrange the scenery, as in Pirandello's *Six Characters In Search of an Author*. The case of radiation theory is unmistakably so: it is more involved, intriguing, and exciting than Pirandello's play. Whether I can retell it so, you will be the judge. But let me try; let me expound reasonably the wrong background to science, and the changes of scenery. True, stuffed-shirt professors do not approve of describing science as a stage of constant upheavals and tumult. If you must agree with them, do judge this book negatively right now and put it aside. Doing so you will save time and avoid the displeasure incurred by the view presented here—of science in flux.

But let me eliminate a false impression I might have given. Stuffed-shirts unfortunately abound. The art of living requires containing and tolerating them. In science, moreover, their chances to do harm are the smallest. Or so we hope. I have no wish to incriminate them but I ask you in advance not to expect this book to boost them. Admittedly one should not make too much of them. I will therefore try to overlook them as much as my narrative permits.

1.4 The Background to Radiation Theory

The big questions for our agenda, let me repeat, are well-known. How do matter and light interact? A bigger question is, what are these made of? The oldest question in the scientific tradition is, what is matter? or, what are things made of? We can similarly ask, what is heat? Or, why are hot bodies so different from cold, and what makes them so? The same can be asked concerning light. Radiation theory relates heat and light. Put a fork in your gas ring and it glows; if your range is electric rather than gas, switch a hot plate on and watch it start glowing as it heats up; if it has a dark spot, see how that spot behaves in the process. These are phenomena which radiation theory investigates.

Not quite. I have now used the word "radiation" in the sense of emission, and so I centered on emission rather than on absorption; and so I forgot all about the greenness of grass, preferring to it some red hot metal.

The question, why is hot metal red? is smaller than the question, why is grass green? Some studies began with the small question, why is hot metal red? As it turned out, it is impossible to study the small question for long without bumping into bigger ones. And even then, we needed better answers to still bigger questions to get ahead. This way radiation theory burst out of its cocoon and appeared as quantum theory, as answers to some big questions, including ones from deep metaphysics, such as, does every event have a cause?

Though radiation theory has the life-span of a century, from *circa* 1800 to exactly 1900, its ancestry is in Newton's book on light of *circa* 1700 (we shall pass quickly over the prehistoric eighteenth century). Newton proposed that light consists of particles, of missiles. The most obvious phenomenon indicating that light consists of waves and not particles is known as Newton's rings (how unfair: though he was the first to describe them with the aid of quantitative formulae, they were described earlier by Robert Boyle and Robert Hooke): put a half ball of glass on a plate of glass and put a light source on top, and you will see lovely rings of color on the plate roughly as if it were a circular pool and someone was dropping pebbles in the middle causing rainbow-like ripples. Newton spoke of fits of easy absorption and fits of easy emission, meaning, presumably, that the disposition of light particles to escape the matter in which they are trapped depends on certain wavelike characteristics. The theory is difficult to follow and unnecessarily complicated. But it raises questions we care about. In particular, where do light particles come from, where do they disappear? Naturally, they come mainly from the sun. The sun is a furnace. What is its fuel? This question troubled him and all his followers, and it was not answered before the advent of nuclear physics, since the sun is a big hydrogen furnace, something like a huge hydrogen bomb. So back to Newton. He groped with the idea that heat releases light captured in matter. But he had scarcely a theory of heat. The important theory of heat which developed in the eighteenth century concerned combustion, and thus chemistry rather than light, phlogistonism and later antiphlogistonism. Both doctrines present heat as a kind of matter. The phlogistonists identified it with phlogiston, i.e., the matter of fire. The antiphlogistonists identified it with calorique or caloric, i.e. the matter of heat; the were materialists, so-called. Arch antiphlogistonist Lavoisier assumed that ordinary free oxygen is a compound of the elements oxygen and caloric, so that when combustion takes place, as oxygen combines with the combustible, caloric is released. What, then, is the source of light? For Lavoisier this is now a most irksome problem since Newton's idea—light is released from hot matter—is not good enough. Is light different from heat? Or is it the same as heat only under different conditions (say, in small units)? Lavoisier tried both answers.

Meanwhile Thomas Young revived an old idea: light is not a kind of matter but waves—waves in a kind of matter. The kind of matter whose vibrations constitute light was called luminoferous ether ("ether" is thin matter and "luminoferous" is light carrying). Now radiation theory had to adjust to the new theory of light. Soon it had to adjust to a new theory of heat. Though the materialist theory of heat was successful, it soon gave way to the kinetic theory of heat that says, heat is motion. In its developed version the theory says, heat is the concentration of energy of motion or of kinetic energy ("kinetic", related to the word "cinema", means, of motion; "energy" means ability to work), and high temperature is high concentration of this kinetic energy. So how does concentrated kinetic energy cause waves in the luminoferous ether? The answer was indicated in the late nineteenth century in a strange way. It is hard for me to describe it here since I have now to show how theories of ordinary matter had changed in the hands of Michael Faraday in the middle of the 19th century. But let me skip Faraday now; I promise to offer a correction later on. Faraday's greatest follower, James Clerk Maxwell, developed his revolutionary idea that when electric charges vibrate they cause electromagnetic waves, and that light comprises these waves (so the electromagnetic ether should do: the luminoferous ether can be dismissed). This looked very promising; all we have to do now is to concentrate enough kinetic energy in a piece of matter (i.e., heat it) until some of it is transmitted to the electric charges which every piece of matter possess (or else it could not radiate!); once the electric charge moves about sufficiently, it causes ripples or waves in the ether; these ripples, these waves, caused by shaking electric charges, carry electromagnetic energy; Maxwell proposed that electromagnetic waves and light waves are identical.

This is Maxwell's general idea of radiation; it is central to the most developed stages of radiation theory. This general idea works so well and so impressively, that it took many years before it was noted that in the matter of interaction between matter and radiation it does not work. Indeed, scarcely were researchers accustomed to its revolutionary ideas when they were forced to face another, bigger revolution: Maxwell's theory, though very successful, fails to explain the general and very famous fact that the color of emission relates to the heat of the emitting body, as it treats all wavelengths as equal; and so a new theory was developed: quantum theory. These days it is common knowledge that short waves are more powerful than long ones, as the very short ones, known as x-rays, damage living tissues. It took half-a-century to learn this fact: it was one of the great discoveries of young Albert Einstein of 1905. When he announced it leading researchers found it most incredible, but it appealed to some students of radiation theory all the same because it offered a solution to standing difficulties.

Radiation theory was an effort to account for colors; one thing stood out from the start, and it was that radiation is a heat phenomenon, yet the theory of heat could not help account for colors—partly because it was not clear how the distribution of heat energy, of the quantity of heat, and how its concentration, its temperature, are correlated. The correlation has to be statistical, but how exactly? Maxwell studies thermodynamics, and Boltzmann improved upon his studies, yet it was solved in stages by Planck, Einstein and many others. In retrospect it is easy to locate difficulties related to radiation: as the theory of light and the theory of heat were both not very good, there is no reason that they should harmonize well, yet harmonizing them was the task of that theory. If a jigsaw puzzle has its pieces finely sawn, and if each member of a group of inexpert players makes a simplified and not very accurate copy of one piece, then obviously it will not fit except by remote accident. Attempting to harmonize pieces is so very interesting just because doing this is trying to work back, by guesswork and tests, from the crude replicas to the more delicate originals. The original is not available, at least not yet, but when we piece some crude pieces together, we do get specific misfits, specific difficulties, which we attempt to surmount. Whatever one may say of quantum theory, however displeased one may be with it, one cannot go back to radiation theory; quantum theories of heat, of light, and of electromagnetic radiation, harmonize with each other and with facts and they have made it difficult to view heat, light and electromagnetism separately. The exercise of harmonizing between them belongs to history.

One general point about the metaphor of the jigsaw puzzle is worth mentioning at once: there is always the possibility to squeeze the pieces that do not fit, to force them to fit with each other. People who are defensive about their theories and are unhappy about bad fits are prone to be glad when someone does squeeze the pieces together. Others may wonder if some valuable information was not lost by the use of violence to the pieces of evidence. In advance we do not know; in retrospect things look much more obvious. And so, you should not be surprised to learn that from the viewpoint of the late nineteenth century the difficultly looks hard to place. We can use this fact to explain why both the theory of light and the theory of heat were then not good enough. It is very easy to do so with some hindsight: the comparison of a theory with its predecessor discloses the advantages of the one over the other, advantages which depict the discord between them, of course, and to the advantage of the new, i.e., to the disadvantage of the old.

The theories of light of the last century all presented light as vibrations of the ether, as results of electric vibrations. These theories were tremendously successful in many ways, but they were not designed to explain the emission and absorption of light; this was left to radiation theory to grope with until it broke down, indicating a need for a new theory capable of doing that. As long as theories of electricity treated all wavelengths on a par none could help us understand the fact that as a piece of metal heats up the color it radiates changes from red to blue.

New vistas were opened by the discovery of the electron—by J.J. Thomson in 1897—of a concentrated electric charge in minute mass. The electron was placed in the atom and made to vibrate. Yet the very idea of a distinct particle of matter such as the electron located in the atom amounts to the splitting of the atom, and it took a decade or more before this idea was seriously entertained. Even then the trouble was not over: the view of a material electron radiating in accord with the classical theory sparked an insurmountable difficulty: if matter carries electric charges, then, as its atoms vibrate, so do the electric charges; and if vibrating matter is electrically charges then it constantly radiates, since the theory said that vibrating charges radiate energy; consequently, it must constantly lose energy thereby coming to rest. This means that all matter should rapidly cool down to the minimal temperature, to the absolute zero. This difficulty was solved by Niels Bohr whose theory of 1913, known as the old quantum theory, distinguished between vibrating atoms and the electrons that vibrate in them; the theory ignored the vibrations of atoms, since usually atoms are electrically neutral; within the atom the theory recognized the nucleus that is more or less the atom and the electrons that vibrate around it; and the theory did not allowed vibrating electrons to radiate unless they could discharge large pellets of energy. The old quantum theory had a limited success, but it was a great breakthrough.

As to the theory of heat, no consensus on it was possible till the second half of the nineteenth century; till then there was a stand-off between the theory that heat is matter, the materialist theory as Boyle had called it, and the theory that heat is the motion of the small parts of matter, the kinetic theory, as historians call it:

(1)　the tendency of heat to spread evenly like a fluid supported the materialist theory;
(2)　heating by friction and heat engines supported the kinetic theory.

Most importantly, the materialist theory allowed heat to dwell in an empty container, but not the kinetic theory; the kinetic theory explained the adherence of heat to matter but the materialist theory did not. Neither theory took account of all the familiar facts of the matter.

According to the new theory the heat of a system is the concentration of kinetic energy in it, and the quantity of energy in it is not heat but a quantity of heat, and in the following way. Very little kinetic energy in a small spark will make it very hot, even though a person who touches it will scarcely sense this, just as the addition of a tremendous quantity of energy into a bath full of water at body temperature will raise the temperature so slightly that the person who touches the water scarcely sense the difference; hence, *the sense of heat requires a rapid move of good quantities of heat.* For example, when two bodies are touched, one a conductor of heat one not, both not much hotter (or colder) than the environment, only the conductor will feel hot (or cold).

Here was a lot of work to be done, especially since the only system of concentration of kinetic energy known was the simplest, that of free particles moving around, of a gas. The study of the behavior of gases at different temperatures gained significance. When the study was extended to the behavior of solids at different temperatures, the difference between the radiations of hot solids and gases became important. Radiation theory was to explain this.

Consider the general laws of electromagnetic radiation as well as the general theory of heat (as concentrated kinetic energy), including the law of the conservation of energy: whatever light a hot body emits is in itself a loss of heat; whatever light a body absorbs is heating. All this is a far cry from the diversity of the phenomena of emission and of absorption: something is *a priori* odd about the program as presented here: a system's total energy, its electromagnetic and thermodynamic traits, these are governed by general laws, whereas radiation is intriguing (not in its generality but) in its variety! The world is full of blazing colors! How can abundant facts fit scant laws?

1.5 Unity And Diversity In Nature

One might cry out in despair. This is a most general problem in all science: how can the diversity of phenomena be explained by their conformity to universal laws?

This problem is as old as science. The ancient Greek thinkers faced it. The first Greek physicist, Thales of Miletus, started Greek thought by declaring that all things are the same. This sounds odd, yet we all agree that although one looks different on different days and is in different moods, one still retains one's (personal) identity, that likewise, ice, water and steam are (chemically) identical. Although matter appears in diverse forms, it also retains its identity: there is only one matter, there is (chemically) one element; all other matter is derivative (compounds) of this matter. In brief, Thales claimed that there is one element, and it is water.

I have put Thales' views crudely, hoping to make them lass strange and to elicit some disagreement. There cannot be compounds of one element, I hope to hear you say; we have compounds of hydrogen with oxygen, and of hydrogen with chlorine, but not hydrogen with hydrogen! Whoever heard of such a compound!

And so, after having stated rather poorly the views of Thales of Miletus, I have succeeded in stating, again rather poorly, I am afraid, the view of Parmenides of Elea, one of the deepest and most influential thinkers in western civilization: Parmenides said, since there is really only one element, there can be neither diversity nor change; their perception is but in the world of phenomena; it is a dream, an illusion.

Now, what I have practically put in your mouth in the paragraph before the last is not true. You can see this for yourself right now if, when you finish this very sentence, before I have the chance to tell you what is wrong with the paragraph before last, you read that paragraph carefully again. For example, consider oxygen and ozone; an oxygen molecule consists of two oxygen atoms, and an ozone molecule consists of three oxygen atoms; there are also free atoms of oxygen not combined with anything and not forming any molecule, vicious in their readiness to bite into anything around in order to join with it into a molecule proper! Hydrogen peroxide contains two hydrogen atoms and two oxygen atoms and releases free oxygen in order to become water which has two hydrogen atoms and one oxygen atom, and this is why hydrogen peroxide is such a powerful oxidizer. Ozone is an even more powerful oxidizer.

This is only one of the many ideas which in their general form occurred to the ancient atomists, especially Demokritos of Abdera, when they answered Parmenides' fantastic claim that there is no diversity. Each atom, said Demokritos, is unchangeable, but combinations of atoms can alter. Different atoms may have different shapes and sizes and different atoms can group together in different groupings, etc. This idea of Demokritos has undergone many developments and transformations in many directions. I am bound to stick with the question we started with: how can we explain diversity by unity? Newton was followed by Immanuel Kant, the sharpest philosopher of the Newtonian era, and by Pierre Simon Laplace, the sharpest mathematical physicist of the Newtonian era, and they had a simple solution. Physics includes general or universal laws and specific initial conditions, and the theory tells us what kinds of initial conditions can be operated with; but they are diverse kinds, and may be diversely combined and yield diverse explanations with the aid of the same laws. Sorry for this terse style; let me expand slowly.

Take Copernicus' theory. It is very simple. The sun is the center of the universe; planets circle round it in constant angular velocities (except the moon which circles the earth). Now, the diversity of the phenomena of astronomy should be deduced from these two laws and from initial conditions of a very well specified nature: astronomers must find both distances and angular velocities. But they have to find them for only one instant and all other instants, the whole world of diverse celestial phenomena, eclipses and transits and the precessions, all follow. (The details required are called initial conditions, since in their basis later ones are calculated.) Or later conditions should follow the laws and the initial ones; they do not, and that engaged Kepler; I shall leave it now.

The Copernican case is simple; things become more intriguing when the initial conditions concern not one single item, be it a planet or a cannon-ball, but a class of items; a mechanism or a model, so-called. We assume, somehow, that all grass is green because each blade of grass has the same absorption-emission mechanism. Indeed, we think all green vegetables share the same mechanism—chlorophyll—even though not all green things do.

It is easy to say that all green plants but not all green things share a mechanism. And so this must strike you as arbitrary—unless the mechanism is specified. But what qualifies as a mechanism? what kind of initial conditions go into a mechanism? There is no general answer to this question. The initial conditions or mechanisms, which one uses to explain diversity with the help of a single law, must depend on that law; different laws invite different initial conditions. Which should we follow? We do not know. We try them all, we check all the facts at our disposal. This is what Sir Francis Bacon said. He required that we should begin in ignorance and so have no reason to decide that one fact is more typical than another, that some facts are simpler, etc.: he concluded that we must consider them all. Radiation theory has an abundance of observed facts. Under the influence of Bacon, it was not permitted to discuss the selection of facts to pay special attention to. Consequently researchers had to make their own rules and decide for themselves which facts to pay great attention to, which facts to ignore even at the risk that research will continues on the wrong track. Some researchers stayed on the wrong track which led them nowhere, others were luckier, but all the facts were just too many—they still are. Sooner or later researchers had to admit that Bacon was in error, that we cannot start afresh and we cannot consider all the known facts together.

When light passes through a prism, it decomposes into beams of different colors—just like in a rainbow. It is the phenomenon of refraction: a beam passing through a glass prism or a raindrop changes its direction and this change differs for different colors: they have different levels of refrangibility, so called, different degrees or indices of refraction. The result is called a spectrum. The spectrum of white light is a full rainbow, that of one color is of that color. More precisely, the spectrum of that color is more intense in that color's region than in the region of any other color—as described in a graph with a hump in that color's region. The graph of the intensity of a mixture of two colors may show two humps. A spectrum may be broken by dark lines, i.e., small parts of the spectrum with very low intensity, or it may be discrete, i.e., consist of a few bright lines. Hundreds and thousands of spectral lines still await explanation. This is an enormous task, as yet incomplete.

Now that we know something about spectra, we prefer particular sources of spectral lines. Vacuum tubes such as the neon and argon and sodium lights which clutter the modern cityscape are most useful; they were invented only after the major achievement in the field was made, namely Kirchhoff's law of 1860 to which we are now coming, and their excellence became evident only about twenty years later. Before that electric sparks between electrodes made of different materials—electric sparks or arcs—were used. The choice of sources of light is theory-based: which facts are the more representative is only found out when they are understood.

1.6 Kirchhoff's Law

Can electrodynamics and thermodynamics together explain the diverse color phenomena? What model are we looking for? This was *the* question. Now we know: electromagnetics made this task impossible. It does offer a hint, though. Its correlation of color with frequency informs us that a body emitting a color contains electrons vibrating in its frequency. Also, it informs us that the energy of a vibrating electron depends solely on the frequency of its vibration. Thermodynamics presents the temperatures of systems possessing given energies (this will be discussed below). So the situation looked promising. Both Lorentz and Planck calculated energies of oscillating electric charges before 1900. The theory got into trouble just at that point: where the pieces of a jigsaw puzzle ought to fit perfectly, there a defect may be spotted—not with ease, though.

The promise failed, and it was great disappointment, as Kirchhoff's law had seemed so promising. It is the pivot of radiation theory; studies of radiation between 1810 and 1860 culminated in it or led up to it, and from 1860 to 1900 they went into details of it. The general idea behind it goes back already to Fraunhofer, 1820, if not before, to Prevost and to Leslie, 1800. It is strikingly simple: we know neither the emission mechanism of any single atom, nor its absorption mechanism; correlating the two may be easier than examining each alone. When two receivers are tuned to the same station, we know that both send the same signal regardless of what it is; the positive and negative of any picture may be correlated with the aid of a numbering system, with no reference to detail: whatever a picture is, its positive and negative show the same image, its black and white are reversed.

Take it slowly: a picture is mostly gray. Moreover, its black and white are relative: there are different degrees of blackness and of whiteness. To correlate a picture with its negative, then, a universal numerical scale is required. Declare the black and the white of any given picture the extremes of the scale, call them 0 and 1, and find their true value on the universal scale; the negative of a gray spot is then the complement of that spot between 0 and 1 and so that the sum of the brightness of a spot and of its negative is 1. Absorption, Leslie suggested, is roughly the negative of emission. A black piece of coal emits white light when hot, a green piece of glass emits red light when hot. (The chlorophyll of a heated leaf decomposes long before it is hot enough to emit red light.) This theory is neither precise nor true. Kirchhoff hoped to improve upon it by a general formula that would correlate every absorption mechanism to its associated emission mechanism. For this some sort of universal scale is needed. Kirchhoff's great discovery was the identification of this sacle as the simplest radiation, that of any black body.

Chapter 2

THE BACKGROUND TO RADIATION THEORY

2.1 Flames as Things

The origins of most scientific theories go back to pre-scientific observations and speculations, not barring radiation theory, the theory of emission and absorption of radiation, of the conversion of light theory and heat into each other. The exchange between heat and light was known since people rubbed sticks or hit flint in order to make fire. Blacksmiths from time immemorial used the color of radiation emitted from hot pieces of metal to judge whether they were hot enough for given purposes. Science picked up when Newton declared fire nothing but radiating air. (Incidentally, he did not know about gases and considered air an elastic fluid; gases entered physics after air was viewed a mixture of oxygen and nitrogen whose components are neither a compound nor separated to layers as fluids should be. Later on the flames were considered not gaseous but plasma, which does not fit the view that matter is solid, fluid or gaseous. But we need not go into that: it should suffice to note that Newton deemed flames hot matter.) Newton illustrated his view with a most beautiful experiment which historians of science regularly overlook; quite apart from its immense importance, it is breathtaking in its simplicity. What Newton wanted to illustrate is that what looks like a flame, what looks like an entity, is nothing but a zone of high temperature. A candle's flame hardly looks less of a thing than its wick; a camp-fire exhibits its dancing flames as dancing just because they look like things. Imagine a patch of green grass on a sandy bank. Imagine that someone told you there is no grass there; there is a drift of sand into and out of the patch, and that each grain of sand turns green when it enters the zone and sandy-colored again on its way out. Now this story is not easily believable, and for a good number of correct reasons. It becomes more credible, and even scientific, when we replace the grains of sand with the chemical atoms in the grain. I shall come to the identity of the patch later on, or rather to the reason why we tend to deny its existence all the same. I mention this case so as to stress the incredible aspect of Newton's idea, that flames are not things but high-temperature zones. (To be fair we should note that already an ancient Greek, Heraclitus, had said that flames were not individual things, that fire was a state of change; all flows; or, metaphorically, all is fire: all so-called things keep changing, keep losing their identity. One cannot step into one [and the same] river twice, he said. Hence, said his follower Kratylos, one cannot step into a river even once!)

Newton took a small bottle, filled it with smoke, and placed it inside a flame. If flames are not things but high-temperature zones, then a flame would heat the bottle and the smoke inside it until it had the same temperature as its surroundings and thus became a part of the flame; that is to say, the smoke should soon glow and merge into the fire. And so it does, of course: smoke is carbon dust; when heated it should glow like coal.

Absorption and emission seemed to Newton to be roughly modes of heating and cooling. A black surface absorbs more sunlight than a white one, and heats more as a result; a hot body is more likely to emit and thereby cool down. Newton even suggested a mechanism: agitation is really the cause of emission, and heat is one mode of agitation; friction is another; so is percussion and even putrefaction (rotting), even electric sparks, the light of glow-worms, and the shiny eyes of cats and dogs agitate and thereby emit light. One who adopts this viewpoint cannot see fire as a body. What else is a red hot iron than fire? asks Newton. And what else is a burning log than red hot wood? Now it looks as if fire gains as much identity as a piece of red hot iron, until it is noticed that not the iron is fire, but rather its being red hot; fire is a state, not a thing. And to drive this point home Newton reports his experiment with a smoke filled bottle placed in the fire until it becomes red-hot, until it glows and thus merges with the fire.

This impressive experiment becomes unimpressive when one takes for granted that flames are not things but states; in particular, it is commonly argued that things retain some measure of permanence whereas flames are ephemeral, blazing and extinguishing as they do (so that only in modern science were the sun and stars viewed as flames, not in ancient science, nor anywhere else). Perhaps Newton's claim that a red hot piece of iron is a flame is not impressive, because just as a flame can come and go, so does the heat of the piece of iron. Familiarity with an elementary course in chemistry may render the claim even less impressive, perhaps, since one learns in that course—what Newton did not know—that any piece of red hot iron can actually burn, that is to say, combine with oxygen to form rust. It burns slowly, to be sure, even much more slowly than a match or a cigarette. But after all we do not deny that a cigarette's end is set on fire just because it burns more slowly than when thrown into a furnace. If you are well informed, you may say, it is all too obvious that flames are not things but conditions of things, and Newton's experiment only illustrates the obvious; the fact that for nearly one hundred years the very best scientists believed in the existence of the matter of fire, phlogiston, after they knew of Newton's experiment, is only one of those small scandals in the history of science; but Lavoisier has rectified that scandal. So there; everything is in order again and there is no need to fuss much about Newton's experiment.

The view of phlogistonism as a prejudice is regrettable hindsight, even though most histories of science that I have ever read share it. As Thomas S. Kuhn has observed, the current dismissal of an idea once venerated is usually an expression of a lack of imagination which is unnoticeably translated into a condemnation of some great ideas. The same, I suggest, holds for phlogistonism: though false, it is one of the greatest ideas of chemistry ever. The next section ought to illustrate this: it presents difficulties concerning the theory of heat. Many historians consider heat from a commonsense point of view and find it utterly unproblematic. It was the received opinion in the last century that the theory of heat is problematic and this opinion is shared even today by some serious scientists.

2.2   Heat as Substance

A flame, most people say, is obviously a condition of things, a state of things, not really a thing proper; obviously, things have some measure of permanence, but not flames. Admittedly, while looking at a steady flame of a candle we may have a momentary feeling of permanence, but we can snuff out the candle and see the flame vanish; things do not vanish, only states or conditions do. Hence, the candle that vanishes either is not a thing or it does not vanish from the face of the cosmos, only from our field of vision. Compare a flame of a candle with a hot piece of iron and ask, is the candle more like the flame or more like the piece of iron? Surely the piece of iron is more stable then the flame and ought, then, to be considered more of a thing. Yet, due to merely incidental optical differences between the burning iron and the candle, the burning iron is considered one thing, but the burning candle is considered two: a candle plus a flame. Had the flame been not on top of the candle but a part of it, we would not view the flame as a thing apart from the candle; indeed, we do have candles with flames not on top of them but as glowing parts of them, cigarettes, electric light bulbs, and even pieces of red hot live coal (which, after all, is almost the same as red hot iron, as Newton realized), and so, also, the candle's wick.

   This is a widespread commonsense theory, the received opinion concerning identity, which deems a thing the piece of metal and the candle, but not the flame. It is a mixture of two theories; their conjunction is inconsistent yet each of them is in itself highly unsatisfactory. It is common to have a popular mixture of two theories even though it is inconsistent: each of the two is insufficient, and one feels that the truth is between them, without being able to articulate it. The one theory, which may be detected in the previous paragraph, is that things proper are permanent and states of things are not; the other is that being a thing, having an identity, "thingyness", is a matter of degree of permanence. Neither doctrine will do. If only utterly permanent things are things proper, then, the discovery that chemical atoms can disintegrate leaves us with no known permanent things proper; if proper "thingyness" is a matter of degree, then the required minimal degree of permanence is arbitrary and shifty. Let me elaborate.

   First, terminology. A thing proper is what philosophers call substance; and being properly "thingy" is being a substance, being fully permanent. When Thales spoke (or rather, when Aristotle spoke in his name) of the substance of the world, he had permanence in mind as the standard argument indicates: one changes one's appearance and moods while retaining one's identity. Hence, continued identity is permanence, and that which provides it is the substance underlying the changes. In nineteenth-century Europe, as in ancient Greece, permanence was granted exclusively to atoms. Today even this permanence is denied; any piece of matter, according to contemporary physics, consists of smaller building-blocks, elementary particles such as protons, neutrons, electrons, and mesons; these are destructible and transformable to energy and/or to each other. (If they are composed of quarks, then, whatever these are, they too are presumed transient.)

Assume that only atoms are permanent. This conflicts with the commonsense view of our bodies as substantive. Now everyone eats and drinks every day of one's life quantities of solid and liquid matter and breathes quantities of oxygen—foreign at first but some of which soon becomes a part of one's body. The body is like the imaginary patch of grass which is not a patch of grass at all but an area on the beach where material things become green when they enter it for the duration of their presence there! Even the ancients knew that the body is a furnace of sorts except that there is nothing in our bodies to compare with the stable walls of a furnace. Even our skins and skeletons keep altering; furnaces, as much as humans, are burning flames, perhaps with their metallic walls and our skeletons as wicks of sorts! Are furnaces and people more substantive just because they maintain their structures for longer than wicks? If so, will the permanent sacred light in temples and on graves of unknown soldiers and of well-known ones, will these come closer to being things? Is the identity of an infant who dies soon after birth threatened by its tragically brief duration?

I have no intention to try and solve all the problems I pose here: I will limit myself to the discussion of classical physics. After Einstein's equation of matter and energy in 1905 matter seemed to lose all of its substance, yet it is difficult to conceive of the material world as mere process with no substance to be processed, and it is no less difficult to identify anything as substance. The field of study of these questions is metaphysics, more precisely the part of metaphysics known as ontology, the theory of entities (the theory of being, to use the Aristotelian jargon). Today most philosophers and most scientists shun metaphysics, and the faithful few who do engage in metaphysical research tend to obscure matters and thus render their discussions quite problematic. Science needs ontology; physics is now in trouble partly because it has transcended its old ontological theories and has none with which to replace them. When a physical system has a clear ontology, that ontology helps comprehend its specific theories, regardless of whether one agrees with that ontology or not. This is one reason why the study of older theories is easier than the study of contemporary ones, and so there is an extra bonus for the study of physics historically—on the understanding that the old ontology is no longer deemed true, of course.

Historically, Aristotle was the philosopher who viewed the material world as consisting of substance and process (in his jargon, being and becoming), yet his detailed studies center on substance and its qualities; he ignored both structure and process—except here and there in his biological studies—which is surprising, as he conceived the cosmos as the process of growth and aging. It was the phlogistonist chemists who introduced process into science in a large way, on a large scale, unforgettably and irrevocably. Consider in this respect the testimony of Joseph Black. He was one of the greatest eighteenth century chemists; he was first an ardent follower of the phlogiston theory and later he become an ardent antiphlogistonist as he was impressed with the beauty, precision, and convincing force of Lavoisier's experiments. Black never ceased admiring the thinkers who had introduced phlogistonism. He said, phlogistonism is what made chemists think in terms of chemical processes. This same assessment was repeated verbatim in the 9th edition of the celebrated *Encyclopedia Britannica* (Art. *Chemistry*). This verdict was altered, I do not know when and how; most histories of science I have read depict phlogistonism as an Aristotelian prejudice. Historically this is preposterous: the chemists of the age, especially the leading phlogistonists, such as Stahl, were obviously adherents to the mechanical philosophy. Moreover, the chemists who were supposed to ascribe to phlogiston negative weight (as Aristotle did) , Priestley in particular, denied this explicitly. But, and most seriously, the error is a failure to comprehend the phlogistonists' concern. What was significant for them was not so much that there existed matter of fire (phlogiston); and they did not identify the element phlogiston with the Aristotelian element fire; they were so impressed with Stahl's ability to describe theoretically and illustrate experimentally many chains of chemical changes, each beginning and ending with the same chemical materials, and each step consisting of transfer of some chemical elements, including phlogiston. As for fire itself, they saw it as a process, of course; they all had read Newton's popular *Opticks*. The process they studied was the release of phlogiston (from its chemical bonds)—and the fire caused by its rapid release. On this Lavoisier had no improvement to offer; for him fire was the rapid release of caloric (from its chemical bonds). What we call today oxygen was, in his view, a compound of oxygen and caloric (both terms were his own invention) which tends to release its oxygen to combine with the combustible matter and the caloric to produce heat in the process which is known as combustion when it is rapid.

2.3 Radiant Heat

What is the nature of light? Is it a wave phenomenon? Does it consist of corpuscles? is it something in between? Recent discussions concerning light are cast in a framework so different from the customary, traditional one that confusion is likely to arise in readers' minds. I found scarcely a history of physics which does not reinforce this confusion. Political scientists warn the public regularly not to mix up instances of one institution (such as a trade union or a house of representatives) of different countries or instances of one institution in one place but in different points of time. In physics, such warnings are unnecessary; though light in the Far East is different from light in the Far West, and it differs most radically in the Far North, the laws governing it are the same everywhere, and they do not change in time. Even scientific theorizing, which is more human and thus more liable to human vagaries, knows no national boundaries and is indifferent to political regimes and cultural depravity. By a universal convention, where national differences enter scientific disagreements, be those in politics, biology, or physics, then at least one party, perhaps both, must be considered unscientific. Of course, in such cases one may ask, who is scientific and who is unscientific, us or them? The answer always is this: we are the scientists and they are the pseudo-scientists, of course. How uninteresting! Whoever is scientific and whoever is unscientific, both parties are on guard, and they know that information and ideas cannot move from one framework to another without recourse to reinterpretation. Reinterpretation is the universal requirement of all travel of ideas (in space as well as in time), political, scientific and pseudo-scientific, in exchange between scientists or between pseudo-scientists, between those who happen to agree with each other and between those who do not. Nothing is more misleading than a literal shift of ideas, a shift without interpretation. Histories of science are often confusing precisely because their authors meticulously transcribe salient passages from different historical figures and from different historical texts, totally oblivious to the obvious fact that the authors they quote often disagree—on fundamentals and on details.

When we consider the idea that light consists of particles, it never occurs to us that each light-particle is eternal; whether light consists of particles, waves, or anything else of whatever nature and character, one thing we all take for granted as an almost unshakeable dogma: light can easily be destroyed. When any opaque body absorbs light, we now all assume, the absorbed light is simply annihilated, destroyed, wiped off the face of the universe. It is true that its energy is conserved, as well as its momentum, and perhaps some other of its characteristics or qualities. It is true that the conserved energy of destroyed light may become light again. If light is assumed to be a particle, it is quite possible—though by no means necessary—that a particle of a certain identity, so to speak, be absorbed and then re-emitted. But (inasmuch as we can speak of the identity of any light particle, or even of a light quantum as a particle at all) no one in their senses claims nowadays that there was a continued identity of the light particle, that somehow the light particle was not annihilated but absorbed, stored, and then emitted. Nor do we say that the emitted light particle is the absorbed one reborn. In a distinct sense we can say that the emitted light particle is quite another, different from the one previously absorbed.

Yet if it is assumed that a light particle is an atom, and that atoms are the substance of the material world, then it follows that an absorbed light particle does not vanish, as it cannot be annihilated; it only hides there. In the eighteenth and nineteenth centuries, let me stress, light particles were generally considered atoms, that is, indestructible. Forgetting this fact makes incomprehensible Newton's explanation of the interference characteristics of light, his view of fits of easy absorption and easy emission; light particles, he said, can be trapped in matter or not, depending on certain conditions. They are trapped, not destroyed! Not everyone believed that they exist, of course, but everyone took it for granted that if they exist they are atoms: basically, existence is forever or never; the assumption that they enter the universe and then leave it looked preposterously complicated and arbitrary. This assumption, that light particles can be created and destroyed, was first made by Einstein in 1905 and more boldly by Niels Bohr in 1913. Also, in 1905 Einstein came up with the view that matter and energy are interchangeable, thus destroying the very distinction between matter and process that is so central to the present discussion. Our views concerning light, concerning all atoms and all substance—of the role of physical theory, no less—were then radically altered. If you expect me to present a new and better theory of these things, a new ontology, then I must disappoint you: I have none to offer. So let us return to the old, comfortable ontology.

Consider light to be corpuscular in the old sense—its particles are indestructible. Assume, also, that heat is corpuscular. This raises a problem: how does the absorption of light increase heat? Remember that nowadays the absorption of light is the absorption of energy and momentum; that, whether wave-like or particle-like, light itself does not usually raise the heat of the atom (there is no such thing, really) which absorbs it, the absorption of its energy does. Yet if heat is not energy but a kind of matter, and thus indestructible, then the emission or absorption of light cannot change any quantity of heat.

Perhaps, then, the matter of light is the same as that of heat! Perhaps, however, light-atoms are merely accompanied by heat-atoms. There is no other possible solution, it seems, within the framework which views both heat and light as material—as substantial. But this is an error. Heat, or quantity of heat, seems to disappear in a variety of simple experiments, the simplest and best-known of which is the melting of ice. This is why we wait for the spring sun to melt the snow; this is why we build a fire to melt ice to obtain drinking water: the melting of ice absorbs a lot of heat. Naturally, if heat is a substance, then that quantity of heat absorbed by the melting ice simply cannot leave the universe. What happens is simply that the heat hides in the smaller parts of the water. This hidden heat Joseph Black called "latent heat" (the word "latent" means hidden). Perhaps the absorption of light leads to the release of some latent heat, just as the arrival of a cuckoo's egg leads to the ousting of its poor original inhabitant.

These, then, are three kinds of solution to our problem. The one favoured around 1800 was that of radiant heat; I shall discuss it in the next paragraph; incredibly, it was confirmed in 1800 in a most beautiful experiment performed by Sir William Herschel, the famous astronomer and court organist-composer. He put a simple device known before Newton to a completely new use: he decomposed sunlight with the aid of a prism and showed that most of the heat from the sun centers further away from the center of the color spectrum in the direction of the red. Herschel discovered what we call nowadays infrared light. Herschel, however, would disagree: he discovered radiant heat and its place on the spectrum. Soon William Hyde Wollaston discovered ultraviolet light in the same way.

Why, however, does radiant heat radiate at all? It does not radiate, it fills space, and moves about with the speed of light. Naturally, however, heat tends towards equilibrium—that is, heat always tends naturally to flow from hot bodies to cold bodies. Hence, when a hot body is placed near a cold body, more radiant heat enters the cold body than leaves it and more radiant heat leaves the hot body than enters it. The phenomenon of heating by absorption of radiant heat, in short, is an illusion; in fact what happens is but a mode of transfer of heat towards equilibrium (i.e., from a hot body to a cold body), very similar to conduction (of heat by metals), or convection (i.e.

heat transfer by moving about hot matter, for example, letting hot water flow into a radiator to heat it and make it radiate and heat the living room). Radiant heat looks as if it leaves the sun and hits the earth; in fact it simply travels in all directions and, by the way, helps the sun transfer some of its excess heat to the colder earth. This is the ingenious theory which was invented by Pierre Prevost in about 1790. Prevost is the father of radiation theory proper. Prevost's law is the first central law of radiation theory which was later included in Kirchhoff's law. Let me reformulate Prevost's law, then, somewhat anachronistically: the flow of radiant heat is one and the same as the process of approaching thermal equilibrium: radiant heat equalizes the temperatures of different bodies; radiation is cooling and absorption is heating.

Prevost's law is known to most historians of science familiar with it as Prevost's *law of exchange*. What are we to make of that law now? I do not mean what role it played in history (I shall come to discuss this soon). I mean, is it true? It seems, at once, both true and untrue: it confuses us. On the one hand, both heat and light are forms of energy and so, naturally, they can convert to one another; and their conversion in a given system into each other occurs in a manner that tends to bring that system to equilibrium. On the other hand, there are other forms of energy around for heat and light to convert into, and in their presence the law need not hold; for example, a body can absorb light without heating up, if and when the heat energy absorbed causes not heating up but some chemical changes which are more conducive to equilibrium; thus green leaves absorb sunlight without heating up.

It turns out that the situation is much less easy to sort out than it looks at first. In particular, today we may confine Prevost's law of exchange to cases involving transformation of no other forms of energy except heat and light into each other. Indeed, the rest of this study will be narrowed to this case whenever possible. This case is so very important that a name for it has been coined: the name given to the exchange between heat and light involving no other form of energy except heat and light is "thermal radiation".

Please note: thermal radiation is not just the radiation of hot bodies; it is the emission and also the absorption of radiation without any accompanying process except temperature change; it is the process of conversion of heat energy and light energy. You can see at once that the term "thermal radiation" could not be invented before the laws of conversions of energy forms were studied, that is, not earlier than 1740; Prevost had published his first study a full half of a century before. By 1860 or so his law of exchange was rather obvious, as was the permission to ignore all forms of energy when studying radiation theory and attempting to limit its study to thermal radiation only; by then it was clear that once thermal radiation was successfully explained, other factors could easily be added, such as the conversion of light energy to chemical energy. But the fact that later studies made Prevost's study obvious should not obscure the fact that he was the pioneer who helped develop the studies that later made his study obvious.

Prevost's law of exchange, then, as well as Kirchhoff's radiation law which came to replace it as the law of thermal radiation, is the law taking care of all and only thermal radiation. It seems that this takes care of a lot of trouble. Light absorbed by a green leaf may cause in the absorbing body not a rise in temperature but a change in chemical constitution (photosynthesis). Or, on the contrary, our troublesome firefly does not cool when it radiates, but burns up in a cold flame some organic material. These facts should trouble us no longer: we try to study purely thermal radiation and ignore the rest. This, however, is possibly a bit of cheating. We do not have, you remember, any *a priori* knowledge to help us decide which process is of thermal radiation and which is not. This, you may also remember, is not very troublesome: a researcher will try once this hypothesis, once the other, and see what result is more interesting. But at times matters are left unattended, and then what is a historian to do? Consider for example the light whose absorption breaks up the crystalline structure of the absorbing body such as that of a piece of ice left to melt in the sun. Does this absorption raise the temperature of the ice? No, it does not. Is it heat? Latent heat was the name for the quantity of heat needed to melt a solid; is the energy which has gone into the melting still heat energy? In other words, what is at stake here, a temperature or a quantity of heat? Is the quantity of heat reflected only in the kinetic energy of the motion of particles or also in the energy which binds them to the crystalline grid?

Consider for another example a beam of light which hits an electron bound to some piece of matter and releases it from its bond. This is called the photoelectric effect, discovered at the end of the last century by a few researchers and made clear by the young Einstein in 1905, who followed Planck's work of 1900, which followed Kirchhoff's law of 1860, which followed Prevost's laws and other discoveries. Now, is this a kind of heating since the electron gets ejected and so has kinetic energy? Or is breaking an electric bond something that is partly kinetic and partly electric? I honestly do not know the answer. I do not think anybody does, and, what is more, I do not think anybody cares, and what is still more important, I think the neglect of this question is quite kosher.

This is, in my opinion, a point of profound interest to anybody concerned with scientific method. It illustrates the fact that at times questions are left in mid-air because they become more messy and less interesting than they used to be and other questions steal from them the attention of able researchers. A historian may ask, what, then, is the status of an old question which has been finally left dangling in mid-air? I do not know of any history of science that deals with this matter, namely with the fact that questions may be (rightly) left in mid-study unattended. I do not know of any philosopher of science who has studied the question of the status of such cases. I found a statement of the fact of the matter, namely, that questions are often (rightly) neglected in mid-study, only in one place, in the remarkable and trail-blazing work of the philosopher and historian of mathematics Imre Lakatos. He did not go further than observing the fact.

The question I was addressing is, is Prevost's law true for thermal radiation? We can similarly ask, is Kirchhoff's law true for thermal radiation? (If the answers are no and yes, then we can view the latter as a refinement of the former, perhaps,) I do not know, as too much depends on the meaning of "thermal radiation" and I, for one, cannot tell what exactly this is. The most natural move is, of course, to define "thermal radiation" as the case which involves only a body's emission and absorption of radiation, leading respectively to its cooling and heating or, still better, to define "thermal radiation" as the case for which Kirchhoff's law is true. This move is the easiest as it will make the law logically true for thermal radiation. I shall take up this point after my presentation of Kirchhoff's law, especially since this attitude is advocated in some important present-day texts. Back to history.

2.4 The Place of Prevost's Law in History

It is easy to underrate Prevost's law. Newton had said already that heat causes emission and he even said that this is due to agitation. Would it not have been a shortcut if heat were identified with agitation? But then Newton noticed other cases of agitation involving no heat, such as fireflies. In Kirchhoff's time all heat was known to involve agitation, thermodynamic energy; all thermodynamic energy is the concentration of kinetic energy, and all radiation and absorption is cooling and heating. But half of this is Prevost's: all radiation and absorption is cooling and heating. This raises the question, what did Prevost, Kirchhoff, and other radiation theorists, do with the cases of cold radiation mentioned by Newton, and including putrefaction, fireflies, and eyes of cats and dogs which shine by night?

They ignored them. Most science writers, historians of science, popularizers of science, and philosophers of science, bluntly brand such conduct unscientific. There is nothing more unscientific, they repeatedly admonish us, than to ignore phenomena which do not fit our preconceived schemes. It was Sir Francis Bacon who said that the error of answering big questions first is that the answer may be erroneous; better prepare the ground by answering the small questions first. Now, why does it matter so much if the answer to the big questions is erroneous? Could we not correct our answers to the big questions by trying to apply them to smaller questions and alter them if we fail in such applications? No, said Bacon. Once one has answered the big question, one falls in love with one's answer and blinds oneself to its shortcomings. It is too much to expect anyone to change views in public—experts will fear ridicule—especially when the views in question concern important matters and their shortcomings are negligible. (One normally tends to flatter oneself, said Bacon, that one's theory is important and the objection to it is not.) Once one fails to pay attention to detail, he declared, one forfeits one's right to consider oneself an honest researcher.

Bacon was a shrewd observer, but he was too cynical. As a matter of biographical record, Bacon's view does not apply to Pierre Prevost. Prevost was remarkable in quite a few respects. He was trained for priesthood, became a lawyer and then an educationist; he wrote on philology, on economics, on the fine arts, and on philosophy; he got interested in science in his thirties through personal contact with Joseph Lagrange, the great mathematician who was a most unusual individual. He became very interested in the works of the philosophical physicist, Alain Le Sage (now regrettably and unjustly ignored by historians of science, especially those who are outside the Francophone world), became his disciple and edited his works to which he added many comments of his own; all along he considered his own work, including the one on radiant heat, a continuation of that of his master, Le Sage. Yet he was open-minded and worked with open-minded people, freely and honorably exchanging ideas and criticizing each other. About twenty years after he first published his law of radiation, he published a book on the same topic (radiant heat) taking account of the variety of views about the nature of heat and the nature of interaction between heat and light (especially light as a release of qualities previously latent; see previous section). And he undertook to restate his law in the least prejudicial manner possible to the large issues of the nature of heat and light. I do not think he could have arrived at his law without first pondering the big questions, and I do not think it was possible to state his law in a manner not prejudicial to any views on heat and light; on the contrary, it was one view of the nature of heat and light that led him to the law, and quite another view of the nature of heat and light to which his law drove perhaps him (this is an open historical question) and, as we are informed, some other thinkers of the period. Prevost's law changed attitudes to big questions.

Prevost's most important influence cannot be discussed here in all the detail it deserves. I am referring to his idea that uncontroversial matters should be presented in a manner neutral to the controversy that they pertain to so as to help resolve the controversy one way or another. A most important book on heat, as well as on mathematical physics, was the treatise by Joseph Fourier on heat conductivity of 1830, written without preference of one theory of heat over the other, in the hope of helping to decide between them.

Prevost's law does not depend on any answer to the bigger questions, but is presented as a challenge to decide which answer is true. But what about smaller questions? How did Prevost take account of the cold radiations, such as those accompanying putrefaction, fireflies, and cats' eyes, which were mentioned by Newton? Historians of science who think it is unethical to ignore facts not in accord with one's theory, either ignore the fact not in accord with their (Baconian) theory that some top-notch researchers have made some great contributions to science while ignoring some facts, or else they severely censure these researchers for their conduct regardless of their success. (Einstein's scientific autobiography presents scientists as opportunists.) A historian of science has no way to study Prevost's work, except to ignore his failings, to censure him for them or to give up Bacon's cynical doctrine of prejudice. For Prevost did fail to take account of putrefaction and fireflies, cases noted by Newton already and hence cases which he knew yet ignored. There are also electric sparks and of cats' eyes. We are nowadays definitely and emphatically of the opinion that not all radiation is thermal (i.e., the conversion of thermal energy). Definitely not all excess of a body's emission over its absorption is due to its being hotter than its environment; putrefaction and fireflies are conspicuous examples, but electric sparks and cats' eyes will not be accepted as legitimate examples of radiation that is not thermal.

Cats' eyes are more problematic than one thinks. Prevost suggested that their glitter is a kind of reflection. He did not know what kind. The case of animals' eyes reflecting in the dark has been studied after Prevost published his works, and the first satisfactory theory of this case, which indeed explains the phenomena, was offered by the young brilliant polymath, Hermann von Helmholtz after Prevost's death. He succeeded in showing how the glitter can be a reflection, and this led him to the invention of the ophthalmoscope, the instrument enabling one to look at the retina (which, indeed, reflects light).

Even electric sparks were problematic. Indeed, they were the more obviously problematic and so thinkers worried about them already in the early nineteenth century. It was Faraday who solved the problem by arguing that electric sparks are nothing else but electric currents. I shall discuss Faraday later on, and then I shall explain why it was so hard to accept—even to take seriously—any of his ideas. Even years after his death, his closest younger associate, John Tyndall, wrote in his *Faraday as a Discoverer* that he could not understand Faraday's ideas, particularly Faraday's view of the current. Yet, somehow, in the eighteen fifties already, in Faraday's own lifetime, his view of electric sparks as electric currents was endorsed by leading radiation theorists like Ångström in order to be able to view sparks as phenomena of thermal radiation. There is a moral to these stories which is obvious, but which should be drawn out explicitly, because, unlikely as it may seem, some philosophers of science or historians of science may dip into this book, and then they may draw a moral from what is said here that is not intended. The moral I intend to draw is therefore better explicitly stated, and as the controversial thesis that it is.

The moral, then, is this. Evidence against theories, against answers to any questions, big or small, is all too often simply abundant. Not always, of course, as we may have the exception, at least, of the true theories (there is no counter-evidence to truths). If you once chance to read an old good science book, you will not fail to notice the abundance of evidence against many of the claims which it contains. But if you read a modern science book, you will not see that so easily or else you can consider yourself a brilliant scientist (and if you publish your results you will, sooner or later, be recognized as such). Bacon was wise and nearly right in observing that we are blind to evidence against our own views. But his contemporary, Galileo Galilei, was even wiser and more nearly right; he said that with great effort some clear-cut evidence may be found against our views, and that this kind of evidence should be taken extremely seriously. This message was repeated in the last century by William Whewell. Sir Karl Popper went further and declared that empirical science feeds on this kind of evidence. Most of his colleagues do not believe he means what he says; some say, he only speaks so in order to drive home Bacon's homily against dogmatism: he exaggerates for a good cause. He does not exaggerate, and he opposes Bacon. We all harbor some prejudices, he says, even the best of us, and science helps us eliminate only some of them; research makes some prejudices explicit and testable and thereby issues challenges to test them; with luck and effort they may sometimes be refuted.

How could Prevost know that cats' eyes do not provide criticism of his views but fireflies do? How could he know why fireflies glimmer? Today we view fireflies as emitting cold radiation, thereby exempting them from Prevost's law. We know these facts because Prevost has published his law. The law challenged people to examine cold radiation, including fireflies and fluorescence; it challenged people to develop radiation theory, thermodynamics, and other ideas, without the aid of which we would still be unable to observe the fact that fireflies emit cold radiation. It led them to restrict the law to thermal radiation, thereby rendering the radiation of fireflies not in conflict with the law, and perhaps also absolving fluorescence. Also, many other historically unrelated developments (or rather loosely related; in the widest sense necessarily all developments are related) made essential contributions to the ability to learn about the light of fireflies, such as certain branches of organic chemistry.

Thus, it is even a merit to overlook obscure phenomena, that is to say phenomena which we cannot as yet study in sufficient detail so as to show that they are decidedly valid or decidedly invalid criticisms of current views, as the case may be. What is required of a researcher in such cases is to be honest and admit oversight, perhaps also to explain it. This, we remember, Prevost did very well. His idea that all radiation is thermal is decidedly false, but his careful study of it and his presentation of its relative merit constitute a great step forward. It made the study of the interaction between light and heat even more exciting, and its domain of application—emission and absorption—much more definite.

There is a new idea in Prevost's law of exchange that has by now become so popular and pervasive that it is difficult to notice its importance. It is the idea of thermal equilibrium.

Today we speak of equilibrium in a very broad sense. The narrow sense is static: a system is in equilibrium, or it is stable if and only if any small change in any of its parts, caused by an outside disturbance, will disappear once the disturbance is gone. Consider a small ball in a circular bowl, or an elastic body: an added temporary force causes displacement, but only temporarily. The extended sense of equilibrium includes the idea from economic theory, of stable systems which grow, and keep growing, undeterred by small obstacles. This is called "dynamic equilibrium". In physics the idea of dynamic equilibrium is of no use. Rather, the idea introduced by Prevost was that of the way in which a system moves towards thermal equilibrium, and this required study: thermodynamic systems may move in different ways to thermal equilibrium, through conduction, through convection (the transfer of hot material), radiation, chemical processes, electric ones and so on! (In elementary textbooks on heat transfer only conduction, convection and radiation are considered, even though most of the heat transfer effected by humans today is accomplished by the transfer of fossil fuel, which is a transfer in space-time through chemical gateways!) Which way is chosen and how? How does a system choose how to move towards equilibrium?

This question is most important in science as well as in technology. Prevost began by observing that at times systems move towards equilibrium by radiating, and his idea had great consequences. But these were slow to come. In particular, radiation repeatedly bothered people and they repeatedly bumped into Prevost's law which required the new concept of thermal equilibrium. Meanwhile, other developments made the field more exciting, such as the development of both the theory of heat and the theory of light, as well as a new and initially unrelated science—spectroscopy.

Chapter 3

THE RISE OF SPECTROSCOPY

3.1 Spectral Lines

Spectroscopy is nowadays a very respectable and useful branch of physics (without which astrophysics and quantum theory would be impossible); it is also widely used in industry. The word "spectrum" means image, and it is here the image of the colors of a given light ray, the image of the different wavelengths it contains presented separately from each other; a spectroscope, then, is a separator of the different colors or wavelengths of a given light.

A spectroscope can be a fancy instrument, but it can be so very much simpler than, say, the oscilloscope that belongs to repair kits of many instruments; simpler even than ordinary electric meters. Learning to operate it quicker than learning to operate a plain camera. However fancy a spectroscope is, it is, essentially, three items:
(1) The separator of the colors of a given source of light: it transmits the different colors separately.
(2) The receiver that records the intensity of the different colors.
(3) An obstacle may be added, a filter through which the light passes, between the separator to the receiver.
Let me repeat with slightly more details.
(1) At the source light goes through a glass prism or a diffraction grating (= a flat piece of glass with fine close scratches on it) through which light coming from an external source passes, so that white light turns into a rainbow; if the source of light is red, for example, the separator will be a rainbow with the violet (almost) missing.
(2) At the receiver there is a screen upon which a photographic plate is fixed, and the darkening of its different parts is the measure of the intensity of the light that falls on it; a scale of wavelengths may be added to the plate; some other light sensitive recording apparatus may replace the film.
(3) In between there may be an obstacle, a filter which may be a film or a thin container of some fluid.

The spectroscope helps analyze either the source of light or the optical properties of a filter. The spectrum of the source is *emission spectrum* and the spectrum of the filter which the light may go through is *absorption spectrum*. Consider emission first. If a given source emits a light of a given observed color, the observed color is scarcely ever its real color: the real color is usually a mixture, and the spectroscope shows quite easily the distribution of light among the different wavelengths. The observed compromise color depends on the relative intensity of the components. The observed color of a given source is often near the color the source emits with maximal intensity, called its maximal wavelength.

One way to measure relative intensities, incidentally, is by using different exposure times and comparing the darkening of photographic plates; the human eye is a remarkably sensitive instrument of comparison. Take a series of exposures of one part of the spectrum with different exposure-times and you can easily compare it to a given single exposure of a different part of the spectrum to determine which one of the series equals the single exposure. Alternatives are automatic instruments that absorb light and convert it into measurable quantities of heat or electricity.

Absorption spectra show what portions of white light are absorbed by a given filter. Thus a yellow filter may absorb all of the white light except for the yellow part, or it may absorb red and blue and thus the light that goes through it appear red. In both cases, the idea is to slice up the spectrum and examine the components of the light a source emits or a filter absorbs.

How does one slice up a part of the spectrum? The spectrum may be continuous or discrete, as in hot metal and in neon light respectively. If it is discrete, we may assume that the light in question is a mixture of wavelengths which can be separated, and then the intensity of each of them can be measured separately. It is impossible to take a single wavelength of a continuous spectrum and examine its intensity. When taking a slice of a continuous spectrum, then, we may slice it thinner or thicker; the question then becomes, how thick should slices be?

This is a matter of decision, just the way we decide to measure age by years, or by months, or by days: when we try to find out the level of anything dependent on age, the thinner the time-slices, the more accurate our measurement can be. When we want to measure the intensity of continuous light, then we do take slices of wavelength, and say of a piece of red hot metal that the intensity of the light it emits in the red region is the greatest, or we can chop the red region to smaller ones and pinpoint the peak of its intensity more accurately.

Some spectra consist of discrete wavelengths, and some spectra are continuous with a few discrete lines (a few wavelengths) missing—and these, the discrete lines and the missing lines, are like positives and negatives; they are emission spectra and absorption spectra, respectively.

Now a good intuition or a good background in either science or philosophy should elicit a disbelief in the previous paragraph *a priori*. There are no discrete spectral lines simply because any film on which a line appears, as assuredly as any line on a blackboard or on a printed page, is a crude replica of a line. We do not need Aristotle to tell us that no real material line is length with no width at all as described in geometry texts.

The precision with which today a so-called spectral line can be determined is something hardly imaginable even to a professional physicist of only a generation ago. It is like taking a very thin sheet of paper, and looking over its edge with a powerful microscope, finding in it so many ups and downs, and some downs so deep as to declare the very thin sheet of paper as consisting really of more than one sheet, perhaps slightly merged deep down in the middle of the page.

Widening a spectral line is easy, of course. Even people with little knowledge of photography can "soften" lines or even take them completely out of a picture by deliberately placing them out of focus; in a picture poorly focussed all contour may be lost; a poorly focussed group photograph may appear as a set of blobs. Similarly, a poorly adjusted spectroscope depicts no spectral lines. To capture any line, the instrument must be sharply focussed (in a manner described in the next section). Though totally erased contours are easy to achieve, totally sharp ones are impossible to achieve: even after the sharpest focussing, the contour lines are not quite lines; when spectral lines are sharpened to the utmost, they get a certain minimal width, already known about a century ago, and called "the natural width of a spectral line".

Let me now repeat the description of the spectroscope more abstractly and accurately. It is an instrument that has a source of light on one side and its image on the other and in between some tool, a prism or anything else, to separate the colors: a pinhole well-lit from the outside may be projected onto a screen inside a dark place, and if the light beam is chromatically separated on the way, then different pictures of the pinhole obtain on the screen, each of a different color, ideally each of them monochromatic. A bigger pinhole blurs the image; for a bigger aperture (so as to let in more light) without blurring one uses a narrow slit. As the spectrum is a multiple image of the source of light, when the source is a slit, the image is not a line but a ribbon; when the light in question is discrete, a bright slice of the ribbon is a spectral line. Historically, the first triumph was the sight of lines in a sufficiently focussed instrument; soon after they only played a small role; they gained importance before Planck entered the scene, but became central to atomic physics only in the nineteen twenties.

This spectroscopic information will do for our purposes.

3.2   The Discovery of Spectral Lines: A Problem

Newton performed experiments which we might call spectroscopic but we do not, even though they were spectroscopic, since they were not detailed enough to exhibit discrete spectra. Since he observed sunlight, he would have observed the dark lines in the solar spectrum had he tried to focus his instrument. He did not. Most historians of science—practically all of them—tell us nothing about the development of the instrument from the days of Newton to the days of Joseph Fraunhofer, the great spectroscopist whom we usually view as the father of spectroscopy. However crude Fraunhofer's instrument was, it was more or less the same spectroscope as the most fancy ones we have today: it had a telescope and a prism or a grating and it showed spectral lines. We still call them today Fraunhofer's lines. So how come Newton's spectroscope did not show lines and Fraunhofer's did?

Whatever is or is not known about the transition, the difference is clear: the latter was well focussed, the former was not. This simple fact has many historians of science rhapsodizing about precision and its immense benefits. The moral of the story, as so many historians and philosophers of science would have it, is quite simple: put a telescope on every instrument you have; make each and every measurement a thousandfold more precise! This moral is positively appalling. A telescope is anyhow not an essential part of a spectroscope. It is needed only for the spectra of heavenly bodies. Although Fraunhofer used a telescope and although before him Wollaston used his own retina for a screen and a special prism, a spectroscope can be made in a much simpler fashion. Let me quote Ernst Mach. I do so for the following reason. He naively expressed surprise at the absence of spectroscopy prior to the nineteenth century, rather than try to explain this fact. Had he tried, he might have succeeded: the significant historical information needed to dispel his surprise was available to him but he chose not to use it. Here is the quote.

"It is remarkable that Newton had not noticed the fixed spectral lines; this would have been easily conceivable if he had experimented only with large circular apertures, but he not only recommended, but actually used, a small rectilinear slit. The only conclusion which can be drawn is that he and his followers did not work with all the [available] requisite conditions accurately fulfilled at the same time, namely, a sufficiently narrow slit, minimum [angular] deviation, and insertion of the screen at the correct distance of the image [from the prism]. The fixed lines are observable without any refined apparatus ... Newton's followers could not have varied his experiments a great deal, or the lines would have been noticed."

Mach makes here two points, one technical and one philosophical. The technical point is simple: a telescope only magnifies spectral lines which are visible anyhow, and they are visible provided the instrument is focussed (and if the image is of a heavenly body, then for this a telescope is needed). Spectral lines are visible when the instrument is in focus: sharp slit, proper angles, proper distance of the screen. The philosophical point is one which Mach shared with his less learned colleagues: experimenters have to try and probe in all directions as results are not known in advance. This view is somewhat anti-intellectualist; it is an anti-intellectual view particularly of science—though, I have to concede, it is quite common even among dedicated scientists, not to say among philosophers and historians and popularizers of science. The anti-intellectualism in question rests on Bacon's popular judgement of theorizing without a sufficient empirical basis as misleading and hence as dangerous; he advised empirical researchers to avoid the danger by not theorizing, which renders a dumb experiment better than a thoughtful one.

A very important physicist by the name of Goudsmit, who made an important contribution to quantum theory and who was editor of one of the most prestigious scientific periodicals of the time, had a strong opinion on the matter. I have heard him speak to an international conference of the history of science, where he was talking on some aspect of the recent history of physics to historians of science from all over the world. I remember what he said on the matter verbatim. (I do not know why his talk is not published in the proceedings of that conference.) He said, "Things have got so bad recently, that nowadays experimenters rarely dare perform an experiment without first consulting their psychoanalysts—I mean their theoreticians." And the huge gathering exploded with laughter at the image of the neurotic experimenter, in need of the theoretician's reassurance, who invites theoretical work to enter experimental design thus utterly depriving it of scientific value. This anti-intellectualist idea is expressed in a popular philosophy of science known as strict empiricism as the advice to perform experiments with no theory in mind so as to obtain pure facts untainted by theory. This anti-intellectual idea, though empirically disproved, should not be underrated: it has dominated most of the modern scientific tradition and because there is an important grain of truth in it. Theories do function as blinkers; experimenters who over-trust theories will tend to perform only the observations that theoreticians predict with confidence; this is self-defeating, because what the experimenter has to come up with are new and surprising experimental results. What are new experimental results? What are surprising results? Bacon said, they are experiments whose results have not yet been predicted by any existing scientific theory. Hence, an experimenter should observe first, and think [and consult theoreticians] later. Hence the fashion of anti-intellectualism. Popper said, no; a new experiment to be is not only scientifically unexpected; it is also scientifically counter-expected. (Experience can depend on theory without bias for it: experiment can and should be designed with bias against the theory which it is supposed to test.)

Let me illustrate this with a historical example. In 1830 Charles Babbage, the famous designer of calculating machines, published an important volume called *Reflections on the Decline of Science in England.* In the *Conclusion* to that volume he wrote character-sketches of two of the most famous English scientists in his day, Sir Humphry Davy and William Hyde Wollaston. (Wollaston, as I shall relate, was the first to discover spectral lines.) While discussing Wollaston he offers a lovely narrative. Wollaston was remarkable in a few senses, one of which was the sharpness of his senses. He was interested in optical and acoustic physiology and made a few remarkable discoveries, including thresholds of perception and the ranges of the visible and audible spectra and the variability of these thresholds from individual to individual. He was reputed to be able to write microscopic messages on glass without the use of microscopes. His chemical laboratory was very famous as he was a great chemist and even discovered two chemical elements. His laboratory was pocket-size, and he carried it in his pocket, being a gentleman of leisure who thus spent time partly in London, partly in the country. Incidentally, he was also a physician, like Thomas Young with whom he explored physiological optics and whose wave theory of light he was the first to defend; he was also one of the first to defend Dalton's atomism.

Charles Babbage was not a very pleasant individual, as his amusing autobiography amply illustrates. He did not like either Davy or Wollaston, and his *Conclusion* is only seemingly flattering to them. But he objected to the view that Wollaston's scientific excellence was simply due to his incredibly good eyes. This was too debunking. He even denied that Wollaston's eyes were so incredibly good. He illustrated his point that seeing requires brains by a narrative.

When Babbage heard about the discovery of dark lines in the solar spectrum, he got naturally very interested, the narrative goes. He went to ask for the help of his friend, the famous astronomer, Sir John Herschel, incidentally the son of the Herschel we have already met and one of the first great spectroscopists and photographers. Babbage asked Herschel to show him the phenomenon. Herschel teased him. He said, I shall place the instrument so as to render the lines visible, yet you will not be able to sec them; you will see them only after I shall show you where to look. And then, said Herschel, after you will have seen them, you will find it hardly credible that you could have missed them. And, Babbage concludes the little but impressive narrative, Herschel's prediction came true.

For perception psychologists this narrative is today not impressive: they can match it by even more striking examples of oversight. In the history of perception psychology, Babbage's narrative lamentably plays no role though it should have impressed some psychologists. For us here it is as if made to order. Now we even have a historical record to the effect that merely adding a focussed telescope to a spectroscope would not have led to the discovery. We also have the historical record stating that Wollaston's discovery was made with no telescope.

But let us not rejoice in an easy victory; not only is such conduct generally unbecoming, it is particularly out of place here. For, if spectral lines cannot be observed without a search for them, how can one ever expect them to be predicted? Was the discovery a miracle?

Even a miracle will not do. Long before our story begins, in the mid-eighteenth-century, Thomas Melville noted that the yellow of sodium light (the famous yellow resulting from sprinkling any fire with ordinary table salt) is of a degree of refrangibility narrower than the limits Newton had ascribed to the color yellow. In other words, he noticed that sodium yellow does not cover the whole part of the spectrum allotted by Newton to the color yellow. As nobody was willing then to deviate from Newton because of a minute, insignificant observation, it was forgotten.

This, too, may illustrate Bacon's theory that every theory—even Newton's—acts as blinkers and prevent us from seeing facts that disagree with it. Yet, as William Whewell has observed, were Bacon's theory true, Newton's theory of light would never have been overthrown!

The first rebel was Thomas Young who in 1800 revived Huygens' wave theory of light. Two years later his friend and collaborator, William Hyde Wollaston started spectroscopy. Though the rise of the wave theory of light was a major event in the annals of science, and though there is a literature on the matter, there is no explanation of how Young came to his initial idea. Let me try and fill this gap.

3.3 The Discovery of Spectral Lines: The Story

Wollaston discovered spectral lines by simply constructing a very thin beam of light (that is, a narrow slit) and by holding the prism near the eye. The technique was certainly new, but as a technique, no doubt, Wollaston had used part of it before, in experiments conducted with Young; Young, as we would say today, was the major investigator. In the earlier research the task was to measure precisely distances of objects from the eye; now, a small variation of the distance or angle of a prism may be sensitively noticed by placing the prism near the eye since it causes an alteration of the color of the image on the retina. The most important part of Wollaston's technique is the introduction of light through a thin slit rather than through a hole (or through a wide slit) as Newton had done.

Most historians of science regularly ignore questions that bother their readers. A number of works on the history of spectroscopy, including Mach's study which is quite prominent, dodge the question, what made Wollaston, of all people, think of an especially narrow slit? Newton was pleased with large holes or slits in walls, and so were a century of his successors; not Wollaston. Why?

This question is dual: it has a technical aspect and a intellectual aspect. Let us take the technical aspect first. He carefully examined small portions of the spectrum; he then examined them for relative intensity. For both of these tasks he needed a well-focussed spectrum! At first he held specially devised prisms carefully near his eyes for that end but later he devised a simple method which is easier and much nicer. He let the spectrum fall on a movable screen with a thin slit in it. In either case, in order to have the separation of each part of the spectrum from its neighbor as wide as possible, the spectrum must be well-focussed: the ray must be thin. A thin slit is preferable to a thin circular hole because it allows more light to enter without the undesired blurring that an increase of the hole's radius may cause—both since refraction depends on the angle of incidence and because the spectrum is a multiple image of the slit.

This leads to the intellectual aspect of our question; it becomes, then, why did Wollaston wish to examine parts of the spectrum in isolation, why did he compare their intensity, and how come he observed particularly those parts where he found his few—out of hundreds—of spectral lines?

Newton had taught that light comes in seven colors, that the spectrum is divided into seven regions, and that each color spreads over its allotted region. The index of refraction of each color seemed to him not a single number as we have it today, but a couple of numbers, each being the index of refraction of the limit between the given color and its immediate neighbor. So he spoke not of the indices of refraction but of their levels. (He did so distinctly and systematically. To be precise, he presented refraction not as a continuous function but as a step function: each of the seven colors has its own index of refrangibility.) Without this fact it is impossible to understand either Wollaston or Fraunhofer because one cannot possibly sharpen the focussing of spectra, it makes no sense to try to focus them, if refraction is discrete! This solves the puzzle that troubled Mach: Newton did not attempt to focus his apparatus as no theory provided him with any reason to do so.

Why and how did Wollaston discover spectroscopy? He wanted to find the relative intensity of light in the different parts of the solar spectrum. But this is still far-fetched: why did he care about that? We can add to this that while Newton's view of the spectrum as divided to seven was taken for granted, the study Wollaston was attempting was unthinkable. Yet the questions remain, how was Newton's view superseded? and why did Wollaston attempt to study the new possibility of the relative intensity of the spectrum?

The story of the refutation of Newton's view of the spectrum as divided into seven colors is the story of Young's odyssey. From it to Wollaston's study is but a stone's throw.

Thomas Young's story is very odd. He started, as a brilliant young scholar, by writing a very nasty review of a book on acoustics. Someone wrote a defense of the book saying that at least Young ought to have admitted that the author's method of tuning organ-pipes by the use of beats was new. In a reply Young denied that. Still, his attention was drawn to this fact. Beats are low tones, undertones they are called at times, that result from sounding two tones which are very near to each other. (Beats are always easy to identify, though this may require a little training; at times they may even be heard, without any training, as proper, very low tones. These are best heard on records of Tibetan monks singing most beautifully very deep tones achieved by singing in beats.) They are, indeed, interference phenomena. This analogy between light and sound had led Huygens to his wave theory of light. Newton objected, saying that light does not go round walls as sound does. Yet Newton admitted that light, like sound, has a scale of seven tones and some rules of harmony (I do not know which: the matter is not clear to me). Young's imagination was fired: acoustics has seven tones and harmonies, yet infinitely many pitches. If light refracts and diffracts, it must also be of infinitely many pitches!

Young thus came to doubt the existence of the seven colors. Young went further than Newton in distinguishing between colors and color-vision. There are infinitely many physical colors, he said, but only a few observable colors or physiological colors. Now if there are infinitely many colors, and light is corpuscular, then there are infinitely many kinds of light atoms. This is offensive, as the demand is to reduce the wealth of appearances to the smallest numbers of real things. If light consists of waves, however, as Huygens had suggested, then the presence of an infinite variety of their wave-lengths prevents this offense.

Fortunately, Young was a polymath: a physicist, a physician, a physiologist and even a cryptographer (he worked on hieroglyphs); his interests in physical optics and in physiological optics fused nicely. He argued that physiologically there are three or four kinds of colors, but physically infinitely many. This led to a problem; Wollaston tried to solve it in a manner which led him to spectroscopy.

Wollaston's chief interest in spectroscopy was to isolate the visible colors from each other, namely, to seek a crucial experiment between Newton and Young. Since each single wavelength has its own index of refrangibility, then by proper focussing one can isolate different parts of the color-spectrum and avoid that spread of each color to a segment of the spectrum, as required by Newton.

Since physiologically the spread is unavoidable, transitions from one visible color to another may perhaps be noticed by properly focussing the spectrum. And so, when he found spectral lines, he suspected that they are boundaries between colors.

It is an error repeated by most historians of science to blame Newton for two errors he did not commit. Even Mach says that Newton must have noted "the distinction between physical and physiological properties of light" for he could not have failed to "observe that to an infinite number of indefinitely small possible variations in refractive index there corresponded only a finite number of color sensations, but he has left us little from which to draw conclusions as to what is to be considered the relation between the two properties." So says Mach, but he is mistaken. According to Newton, there are not infinitely many refractive indices, but only seven degrees of refrangibility. Second, Newton is blamed for having assumed the existence of infinitely many different kinds of light particles whereas he assumed the existence of only seven kinds.

(Upon reading the previous paragraph the leading Newton scholar Zev Bechler protested. He quoted from Newton's *Opticks* an interesting passage, one that introduces degrees of refrangibility (Book I, 2, ii): speaking of the separation, from "the most refrangible Rays" to "the least refrangible Rays" of the "Series of Colours" of the spectrum, Newton says, that they, "together with all their intermediate Degrees in a continual succession perpetually varying". This is a recognition of the infinite variety of degrees of refrangibility. Perhaps Bechler is right. The target of Young's attack then was not Newton but either Newton misrepresented or Newton modified. But I do not agree with Bechler. For, immediately following Newton's description just quoted I find (Book I, 2, iii) the following: "To define the Refrangibility of the several sorts of homogeneal Light answering to the several Colours", and after describing some experiments he says, "Now these Intervals or Spaces subtending the Differences of the Refractions of the Rays going to the limits of those Colours ... " and then he provides intervals for refrangibility.)

This explains why Newton never concentrated on small portions of the spectrum, and why Babbage could not see darkening of small portions of the spectrum even when they stood before his eyes. The rise of Young's wave theory of light and his physiological optics together gave Wollaston the reason for focussing parts of the spectrum and thus the possibility to observe spectral lines. What Wollaston observed did not fully conform to his expectations. To secure the objectivity of his experiments he repeated them with diverse prisms and light sources. He discovered lines of emission spectra that way which did not always agree with the absorption lines he observed in sunlight. He found his study intriguing but very inconclusive. He therefore could not decide as yet how many colors were there, and given the number of colors, whether they are or are not isolated by spectral lines (since, if there were, then they may help decide that number); and later he joined Young in suggesting that there were infinitely many colors.

And then came Fraunhofer and destroyed the hypothesis that spectral lines are boundaries between colors with the discovery, with the aid of a powerful spectroscope, of "an almost countless number of strong and weak vertical lines." To invoke his dissent from Wollaston he added, "The strongest lines do not in any way mark the limits of the various colors; there is almost always the same color on both sides of a line, and the passage from one color into another cannot be noted." The meaning of all this is clear to anyone familiar with Newton's and Wollaston's spectroscopic work.

With this development, the link between spectroscopy proper and physiological optics proper were severed once and for all. The classical question was, how many colors are there? It is answered in physiological optics by a small number and in physical optics by infinity. Of course, a very important lesson came out of the brief cooperation: it became clear that colors and color-vision are distinct. In principle this was obviously the case, as noted by such thinkers as Demokritos and Galileo, but only early in the nineteenth century was this established empirically. And so the theory of color vision from then centered on discrete colors and the theory of colors centered on the continuous spectrum. Fraunhofer ended the use of discrete spectra for physiological optics and began a new use for discrete spectra and their study, i.e., spectroscopy, the chemical analysis—especially of sun and stars. The idea that the light from the distant stars may tell us about their chemistry is surely very exciting.

3.4 The Discovery of Spectral Lines: A Discussion

Did Wollaston influence Fraunhofer? The most widely accepted view on the matter is, no, and for two reasons. In his original memoir Fraunhofer does not mention Wollaston and declares his primary purpose to have been the determination of indices of refraction. Now, no doubt, Fraunhofer was very much of a technician: his father was an instrument maker, and so was he. And what is more understandable than a technician's desire to improve instruments? Need we explain such a desire any further? Is it not unconditionally commendable?

There is an inner logic, an inner challenge, inner poetry, in every piece of remarkable technological advance. There is much beauty in Fraunhofer's and others' improvement of optical instruments. But this should not prevent historians from explaining a historical episode; saying that scientific progress was assured when so-and-so constructed a new or an improved instrument is not an explanation. Admittedly, at times this happens: new instruments often promote scientific progress tremendously. They may help solve previously unsolved problems and serve as incentives to explore avenues of research previously neglected as too arduous. This is not to encourage indiscriminate investment of efforts in improving and inventing instruments regardless of their possible use. Admittedly some people develop incomprehensibly obsessive desires to develop some techniques or instruments and possibly there is no more to tell about improvements reached this way. The obsession can be magnificent and have admirable results. In such cases this is what a historian should report. Still, not all technological innovators are obsessive. Even were that so, historians can attempt to explain the magnificent result of some obsessive work and how success had evaded other obsessive and talented instrument makers: historians can try to explain in what respect and manner success had come about, what obstacles were overcome, etc. We want to know why some improvements were successful and others not. After all, the successful obsessive innovators were successful just because their efforts had led to worthwhile challenges to researchers to use their innovations. What, then, was the challenge that Fraunhofer faced whether he was obsessive or not? (He was not; obsessive people tend to be difficult, and he was a lovely person if his writing style is any indicator.)

Fraunhofer improved optical instruments; his chief improvement was in the area of achromatism, that is to say, he attempted to eliminate rainbow-like boundaries of images (created by poorly adjusted optical instruments, including color-television sets). Though already Newton eliminated some achromatism, he could have no satisfactory theory of achromatism; such a theory has to be is a corollary to Young's achievement: refraction, refrangibility, being a matter of continuous alterability rather than of merely discrete grading, opens up the study of achromatism with almost immediate applicability to a variety of optical instruments.

So, it seems that Fraunhofer's study depended on Young's study, possibly also on Wollaston's. Why, then, did he not mention Wollaston, or at least Young?

I cannot do justice to this question here. To show how complex this matter is, let me mention the oddest story in matters of acknowledgement. Young had to mention Newton's work, of course; so he contend that he was following Newton, although he was clearly contradicting him. As evidence for his contention Young mentioned observed facts known to Newton which he, Young, considered refutations of (Newton's) corpusclarian view of light.

Young was overawed by the idea of disagreeing with Newton and this is reflected in his acknowledgement. Even Einstein, in all likelihood the most outspoken and brave scientist ever, was overawed by this: "Newton, forgive me!" was the moving expression in his scientific autobiography. There is, generally, no reason for historians to imply, as they often do, the erroneous view that our standards of acknowledgement are identical with those of our predecessors. The chief rule of the scientific community concerning criticism was introduced by Robert Boyle, endorsed by the Royal Society of London in its early days, and reinforced by Newton. It requires that criticism should be implicit: preferably the criticized hypothesis should not be mentioned and definitely its author and/or defenders should not be mentioned as this amounts to picking a quarrel. This rule is stated and defended and explained in various works of Boyle, especially the prefaces to some of his more popular works. A distinct echo of it appears in Ben Franklin's lovely autobiography. This offers historians much detective work. Boyle's proposal was no iron rule: Faraday consciously tried to institute a more friendly attitude to criticism and more openness about it. Yet already a generation or two earlier Prevost took sides cleanly and opposed cleanly; when he put his doctrine as neutrally as possible with respect to the nature of heat he did so not in order to avoid controversy, but in order to place it in the proper context.

Fraunhofer was not as advanced as Prevost. He was as critically minded, as he attacked Wollaston and bravely followed Young, the almost ostracized opponent to

Newtonian optics, but he followed the conventions of presentation of the age. Fairness requires further comments. Fraunhofer had more reason than Prevost to put his empirical finds neutrally. And he did not avoid controversy when he spoke of the index of refraction as something varying with wavelength, as this amounts to concluding that there are infinitely many kinds of light particles unless the wave theory of light is true.

The great and honest historian, Ernst Mach, stresses that Fraunhofer's discovery "was due ... to interest in scientific problems"; he was unintentionally unfair to Wollaston in attributing the discovery solely to Fraunhofer, and he was unfair to Fraunhofer when he identified the scientific problem with one of purely technological significance. But let us not blame Mach: Fraunhofer had asked for it. At the end of his first memoir on the spectral lines we now name after him, Fraunhofer says, "I could, owing to lack of time, pay attention to only those matters which appeared to have a bearing upon practical optics. I could either not touch other questions, or at most not follow them very far." The first sentence sounds unduly apologetic. The second observes that his memoir was seemingly purely experimental (and so unobjectionable). It obviously hints at some obvious conclusions. The text goes on to encourage readers to follow the path which "seems to promise to lead to interesting results ... [so] that skilled investigators should devote attention to it." He did not explain, but his contemporaries could be trusted to see his point. It is no accident that his chief fan, Sir John Herschel, instituted the wave theory of light in England. (Young himself had failed to do so and was practically ostracized.)

To prevent the impression that Fraunhofer was a bit of a coward, let me stress this again: non-commitment and the glossing over of criticism were favored in those days, as evidenced from the works of outspoken, criticallyoriented researchers like Prevost and Fourier. when Fraunhofer wrote noncommittally the wave theory of light had already been established. This seems to me to indicate clearly that he used the theoretically non-committal language as a matter of style, and even demonstratively so—for good reasons and for not so good ones, of course, but in tune with the convention of the day, with the inductive style of presentation. I would love to report that by now this style is antiquated; in a sense this is the case, and at least in theoretical physics this is so; but in many branches of empirical research the inductive style is still common and even imposed by editors. Perhaps Fraunhofer, too, was made to follow the custom.

3.5 The Rise of Astrophysics

The chief discovery in spectroscopy was that of the dark lines of the solar spectrum. Once these were discovered, emission spectral lines immediately turned up as well. The reason is rather simple: once degrees of refrangibility were replaced by indexes of refraction, these apply to emission and to absorption alike. Color flames are well known; red, yellow, green, blue flames—these were all recorded already in Antiquity. The obvious fact is that the light of a color flame passed through a prism has a spectrum that is in part bright and in part dark. Once the spectroscope was focussed enough to obtain the lines, the bright spectrum of a color flame could also be focussed to lines. (The word "lines" is very inadequate. The spectrum is usually given in a two dimensional strip; but its colors do not depend on the width of the strip; the width is quite arbitrary, and all that matters is its length. To be precise we should speak of spectral points along the length of the spectrum; and, to repeat, the points are not quite points since they do have lengths.) Hence, emission spectra should not look surprising since they are merely focussing sections of the spectrum better, perhaps to a much higher degree than could be imagined, but in principle not spectacularly.

This looks like an adequate story of what has happened only as long as we ignore Newton's theory of color, as researchers have done. This theory does not allow a spectral focussing which gets a clear-cut difference between two parts of the spectrum allotted to the same color, yet it did present absorption and emission as the reverse of each other. The same holds for the wave theory of light, as will be explained soon. Thus, the major aspect of spectroscopy is somewhat less surprising than the very existence of spectral lines which are characteristic of the matter emitting them.

Now we have two questions on our hand. First, what to do with these spectra? To what use can they be put? Second, how can one explain them? What is the cause of spectral lines?

The way the two questions are here stated may be surprising. It is not usual to put them this way. It is a significant and powerful part of the tradition of science to seek enlightenment before utility. Indeed, until the twentieth century, this was a point of honor and of rationality. It seems reasonable that understanding inaugurates control over nature (control over nature not due to understanding is either accidental or mysterious). So the order of the questions in the previous paragraph is wrong.

Allow me to draw attention, first, to an obvious fact that all too often escapes notice. The ordinary sense of usefulness is often, but not always, the increase of human control over nature, the increase in the availability of resources and of comfort; in science the sense of usefulness often is something quite different, namely, usefulness in the search for knowledge, usefulness for the advancement of learning. True, the attainment of knowledge brings about quite often, though definitely not always, the attainment of usefulness in the ordinary sense, in the sense of comfort and the control of the environment; nevertheless, for those concerned chiefly with the advancement of learning usefulness is not for comfort but for this advancement.

The above description is of the tradition of classical science. Ever since the revolution in physics at the turn of the last century, the tradition of science has become more and more open to the view of science as an instrument of control over nature—instrumentalism—rather than as an instrument for the advancement of learning—intellectualism. This theory, instrumentalism, is sadly the demotion of science from the loftier intellectual position it once had. Two points, however, should be conceded—one of which I shall use with respect to Fraunhofer now, and one of which I shall reserve for my discussion of Kirchhoff's important priority claims. First, the word "usefulness" is meant here in a more intellectual sense than it might seem; second, the intellectualism of classical science was marred by its claim of having attained finality, absolute certainty. At least, we will see, Kirchhoff's claim for finality was a very important point in his priority dispute.

Fraunhofer was a technician; consequently his technical bent predominated his work. It is particularly important to deny that technicians only care for patents and financial reward. Thus, Fraunhofer wanted to, but could not, put spectroscopy on a theoretical basis (this took a century of ingenious hard work). He could not put it to any technological use in the narrow sense of technology, in the sense of increasing human control over nature. But he immediately put it to use all the same; he started to use spectral lines as a means for the identification of light sources.

You can perform now with ease one of the most fundamental experiments in quantum theory. Go to a kitchen with a gas stove and turn it on. Observe that the flame is blue. Drop on it a few grains of common table salt. Observe the yellow streaks in the fire where the salt fell. It takes some sophistication, not much, to repeat the experiment with a spectroscope. The yellow light turns up as a bright vivid single line—the so-called sodium D line. Focussing the instrument carefully, as already Fraunhofer did, reveals that this is not one line but two; a so-called doublet. (Not all pairs of lines on a spectrum are doublets, however; indeed, most of them are not.) A doublet is a couple of lines emitted from the very same kind of source, and by almost identical parts of the emission mechanism. It is easy nowadays to introduce novices to spectroscopy and draw their attention to the sodium D doublet. It is hard to grasp how much trouble this single (sorry, double) line caused, how much hope it raised and shattered. It was the first spectral line observed in history—long before Wollaston observed spectral lines in the late eighteenth century—which, you remember, was consigned to oblivion as un-Newtonian though it is very bright and hard to overlook. Wollaston saw it, and noticed its resemblance (in position, of course) to a dark line in the solar spectrum. So did Fraunhofer. Was sodium the cause of this vivid line? Was it sodium that made its dark double in the solar spectrum? These questions were raised and re-raised, discussed to and fro, until Kirchhoff settled the matter with the utter finality which, he insisted, was becoming to science. And soon, of course, people claimed he was too high-handed and unfair to his predecessors. He was scornful. There was a question, a yes-or-no question; in reply, some said, yes, some said, no. This requires no effort, of course. But he, only he, demonstrated the truth of the affirmative answer!

The violet mercury line is another prominent spectral line which has as long a history as the yellow sodium line. Dropping mercury on your gas fire in order to see it is not necessary, nor is it useful, as what is needed is mercury vapor; but sufficiently many mercury lamps or arc-lights are available in street lights in every big town, and you must have noticed how violet their light is, especially when trying to observe red things and seeing them as violet-black. (Incidentally, the yellow lights that flood the highways give off the sodium D-line.) This violet was later even more troublesome than sodium yellow was earlier. But let us leave all that.

Fraunhofer could, perhaps, identify one line here, one there; but he saw "an almost uncountable number" of them. Still, he did compare two solar spectra and see a pattern recurring. He could compare the sunlight with moonlight and light from planets and obtain the same recurrent pattern! He thus identified sunlight as reflected from the moon and the different planets in helping to establish facts not so much about astronomy as about spectroscopy. He then took spectra of stars, including Sirius, and got different patterns.

Blessed Imprecision! These were a series of wonderful observations, and they could not be but for imprecision. As any spectroscopist will testify, moonlight is significantly different from sunlight. This difference was carefully studied so as to draw information about the nature of the surface of the moon before the days of the lunar probe. Even Fraunhofer noticed some differences between the light of Venus and sunlight. He explained it away on the technical ground that Venus' light is feeble (and his spectroscope not very accurate to begin with). It is essential to see spectra superficially in order to say that sunlight and moonlight resemble each other and differ from the light of fixed stars. The two are similar only as first approximations, as in truth they do differ. In 1905 the decisive move was made which, when it became a part of the scientific tradition, rendered first approximations respectable and standard tools of research. In that year Einstein presented a theory, the restricted theory of relativity, and showed that the Newtonian formula for the quantity of the kinetic energy of a body is an approximation to the Einsteinian equation of mass and energy. Before Einstein, first approximations were deemed not good enough: Galileo and Newton disapproved of them. Afterwards, few dared speak against Newton's theory on the ground that it is a mere approximation.

The series of observations reported in Fraunhofer's first spectroscopic memoir sufficed to establish spectroscopy. Spectroscopy came with its problems: how are the dark and bright lines related? Also it came with a promise of a powerful technology: the lines are possibly means of identifying the chemistry of their source. And so a new exuberant science sprang into being—astrophysics, the physics and chemistry of stars. It provided clues to the baffling secret of stars, including one of the greatest and most baffling for over two centuries, one which Newton puzzled greatly about, as many before and after him did, one great mystery and miracle of nature: what are the sun's immense resources?

The discovery of spectral lines in the invisible regions of the spectrum—the ultraviolet and infrared—suggest that these rays constitute not radiant heat rather but invisible waves of the luminoferous ether. Young even proved this by finding interference patterns of infrared waves, thus showing that the difference between red and infrared is entirely physiological, not physical. Incidentally, Young's experiment was the first use of photography ever! In his time photography was already well-known, that is to say, the photochemical effect was known and well illustrated with the darkening of silver-chloride by light. Silver-chloride was smeared on one plate of a *camera obscura*, and so a photo was made. But all one could get were ephemeral negatives. Only decades afterwards did a method of fixing a photograph develop, and still later, a method of using latent (hidden) images on negatives to be developed and fixed. But for Young this was not necessary; observing rapidly fading interference patterns was quite enough for his purpose!

To return to Fraunhofer. His researches reinforced Young's view of radiant heat as invisible light, as they treated the visible and invisible spectra on a par; and they raised hopes for the development of the new science of astrophysics. This is what I mean by usefulness in the intellectual sense.

Unquestionably, science does have a concern for usefulness in the ordinary sense. To some degree the researches into emission spectra in the twenties and thirties of the nineteenth century were economically motivated. The great scientific leaders, including Sir David Brewster and Sir John Herschel, were people of both a terrific theoretical bent and an acute sense of technology and of the marketability of scientific ideas—not to mention the public relations of science. The search for monochromatic light must have fascinated them at least in part because they could foresee in it much technological usefulness (as such light can be used for marking chemicals) and they saw in advanced technology the opportunity of providing jobs for the rising class of researchers from humble origins. (The idea of the use of spectroscopy in chemical analysis was particularly appealing as long as sodium, for example, was supposed to emit only its famous yellow D line!) This vision of the immense usefulness of spectroscopy for industry was shared even by Bunsen and Kirchhoff who, between them, discovered and mapped thousands and thousands of spectral lines. This would no doubt astonish a modern quantum physicist who has learned in the elementary course on quanta of ever so many kinds of sodium lines. Nevertheless, the fact is that though sodium, or mercury, emits thousands and thousands of lines, it is the high intensity of one line (sorry, two), the yellow sodium (duplex) line and the violet mercury line, that is so useful in street lighting. Anyway, it was the search for a monochromatic light that led to the idea of the distribution of radiation. If a flame or a spark emits different wavelengths, different spectral lines, where does most of its energy rest? what is the wavelength of maximum intensity? These excellent questions emerged from practically motivated researches, and answers to them are easily available.

The following two questions are the most important in the field; I do not know when they first received their articulation, but I will later argue that they were there, however vaguely, from the very inception of spectroscopy. Here they are. Are dark and bright spectral lines mirror images of each other, and if so, what atomic structure causes it? Alternatively, how do atoms emit and absorb radiation?

Though spectroscopy is full of problems, its history can be written in a smooth manner as that of cases of successful moves with no attention to the problems which these moves came to solve, and with no attention to failed attempts at solving the same problems. Suppose this is inevitable; then readers should be warned. They are not, and this is deliberate: it is the expression of an attitude concerning the public relations of science, according to which the general public should admire scientific achievements, not scientific difficulties. This attitude is distinctly dangerous to science.

I wish to confess, then, that I too find some simplification of history unavoidable, and that I, too, simplify, even though not so much as to pretend that science is purely a success story. I try repeatedly to make science less of a vulgar success story than most historians of science indicate by illustrating the view of my former teacher Karl Popper that in science finding faults is finding a great incentive to improve matters and an indication of the way in which improvement may be sought. Yet clearly not all errors are equally important, and so not all criticism is equally valuable. In order to avoid simplification I should say what error and its subsequent criticism is important and why. This is easier said than done.

Let me illustrate this with a few observations on one very important paper by Ångström. The paper raises many problems, and I shall ignore for a while its main thesis for which it is historically so important. Let me illustrate how misleadingly selective presentations of history can be. Ångström considered the theory that each element has its characteristic spectral lines and that spectral lines are always characteristic of chemical elements. Try hard as he did, Ångström failed to detect spectral lines of carbon and of sulphur. This led him to the conclusion—we shall see later how very deceptive it was—that traces or impurities do not contribute to spectra, that when studying spectra we need not worry at all about the purity of the material under analysis. Moreover, many spectral lines were not quite lines but pairs of wedges with the thin edges only barely indicating lines. Furthermore, he had to subtract from his spectra the spectra of air, or other materials, which willy-nilly participated in the emission of light. Here the discreteness of the spectral lines was of great help, but then he also found portions not so discrete.

Why, he worried, is matter the source of continuous spectra when solid but the source of discrete spectra when gaseous? He gave some answer, qualitative and vague. Remember Newton's query: is red hot iron a flame? Is a burning log merely a piece of red hot wood? The answer is, in a way, yes (both the solid and the gas contain radiating atoms) but in a way, no (usually solids emit continuous spectra). If Ångström knew why, he would have felt more confident that he was not fiddling too much with the data. As it was, the data conflicted with the hypothesis that spectra depend only on the chemicals emitting the light; spectra also depend on the material state—solid, liquid, or gas. As Faraday suspected and Zeeman and Stark proved long after his death, the magnetic and electric conditions of the environment also influence spectra. Yet under all these conditions the emission and the absorption spectra are symmetrical: whatever effects emission effects absorption too.

One more point. Doppler's discovery of the variability of color due to stellar motion was made a short while earlier and looked promising. Ångström suggested that Doppler's effect could be tested spectroscopically as the speed of sparks producing spectral lines must be high. He was relying on Faraday's revolutionary assumption that sparks are electric currents and on Wheatstone's measurement of the speed of currents in conducting wires. Doppler's effect later proved to be of great service to spectroscopy, when the width of spectral lines became important: heat agitates atoms and their agitation causes changes in the color which they emit, and this smears spectral lines: emission from hot bodies involves emission from atoms which move in varying velocities: the color of the radiation they emit changes due to their motion; as their speed is not equal, the changes due to speed are also not equal. Hence, the colder the source of light, the less its emission varies due to the Doppler effect and the sharper the emitted spectral lines.

I shall not return to all this. I have only tried to give you a taste of how different the labyrinth really looks to the pioneers from what it looks like to those armed with their results. No historian of the subject raises the annoying problem, do spectral lines represent elements and nothing else? Does every element have its spectroscopic fingerprint? Historians claim that the answer is yes, yes of course. And they proceed to tell you who was the first to discover which spectral line is the fingerprint of which element. But think of fingerprints proper; remember how long it took before fingerprint techniques were so perfected as to raise no doubt about them as means of sure identification; think of how many means of identification which looked equally promising in the recent past turned out to be not good enough. Think of the scientific duty to doubt! And notice this: the answer yes, yes (of Ångström and of others) is false. The true answer is no. Some spectra are emitted not from atoms but from molecules; hence they depend on chemistry; other spectra depend on physical chemistry and on a lot of other factors; yet some depend on elements and these can serve as their fingerprints.

3.6 Clues and Promises in Science

The present study is not a run-of-the-mill work in the history of science. It is an attempt to exhibit constantly contrasts between ideas of the great masters of the past and those of the great masters of the present, in violation of the demand, which is popular these days, to avoid such contrasts. This demand, you may remember, is based on the pronouncement that it is impossible to contrast two successive answers to one question without misleading (meaning that it is bad public relations for science). As my story of Young's attitude toward Newton illustrates, this demand and this pronouncement are not traditional. Traditionally, it was preferred to tone down differences of opinion as if they were unsavory. Until Alexandre Koyré developed his way of writing a history of science half a century ago, it was not customary to stress researchers' errors. Yet here this is taken for granted and my effort is to illustrate the importance of mistakes and criticisms, not only as means of avoiding condescension towards the great thinkers of the past, but also, and more importantly, as means of comprehending past problems and their solutions. To that end it is important to notice, however, that we differ from these thinkers of the past not only on the nature of things, but also on the nature of science and on its methods of research.

Let me repeat: the value of a scientific idea or experiment seldom increases due to further developments; once deemed valueless it is either forgotten (to be repeated, if necessary) or fought for tooth-and-nail. Usually ideas are overrated so that they too rapidly decline in value. Incidentally, the value of experiments usually declines too. Is it not the case that my last paragraph contradicts all that? Did I not admit that the value of spectroscopy for the fingerprinting of chemical elements was dubious but nevertheless became widely accepted? Is this not an increase in value?

In a way it is. And there are many similar examples. When Max Planck chopped up the energy of a ray of light, he did not dream he was beginning a revolution: it was hard enough for him to sell his idea even without making big claims for it. This and similar examples have led I. Bernard Cohen, the justly distinguished contemporary historian of science, to say, one seldom evaluates one's own contributions correctly.

Can I reconcile my view with that of Cohen or need I disagree with him? I should like to disagree with him (in a fruitful manner), and I hope I can at least draw attention to a difference in tendency. The actual statements, however, need more clarification before nailing down possible disagreement.

When we care to dwell upon past ideas, in our capacity as historians, as scientists, or otherwise, these ideas look to us far richer in consequences than they looked to our predecessors. We know more about consequences from every theory, both deductively and historically, than its originator possibly could. Now, of all the ideas which we may care to dwell upon, we tend to dwell on those rich with valuable consequences. And so, due to intellectual progress, looking back we see that ideas have produced much higher a yield than initially expected. Just as Cohen claims. Yet of necessity we are highly selective. It is therefore useful to formulate an adequate and clear idea of why we value any item. There are at least two distinct and different modes of evaluation: with and without historical perspective, with and without hindsight. Historical perspectives change, and evaluations based on them may change too, for the better, we hope, since hindsight is meant to improve judgment. Contemporary evaluation is, however, no less interesting. We may ask, as historians, not only what was the contemporary evaluation, but also what should it have been? This is no idle question; we answer it and we do so carelessly when we carelessly and unfairly judge all of Planck's contemporaries to have been stuffed-shirts. We hope we can learn from past mistakes! But when we ask what should they have thought of him, we must, in all fairness, attempt to avoid all hindsight since hindsight may easily provide arguments which were not accessible to them! (Zev Bechler objects; he thinks some hindsight is useful. I daresay he is right: we do not have a theory of hindsight; the admonition of political historians against it are baseless. For my part, I try to avoid the worst kind of hindsight, the one so characteristic of so many historians of science.)

We should not rush to condemn Plank's contemporaries, for example. It is no use to say, the immense success Planck's ideas had in 1913 or in 1926, shows that his peers ought to have valued them in 1900; complaints about the lack of foresight are unenlightening. How can one evaluate a contemporary evaluation of a scientific item without using hindsight, i.e., without ascribing any foresight?

It is this question that Sir Francis Bacon has addressed. His answer is shrewd and provoking, yet it is impracticable. Bacon said, we have no ability to predict scientifically the future scientific use of any scientific item. To predict this conjecturally would be prejudicial. Hence all scientific items should be valued as potentially equally useful, store them, and hope for the best. This advice is what makes Bacon's answer impracticable. Most items of information that are compiled with no eye to future use are ignored almost at once, and rightly so. We do have an idea about how promising an item is right now, and our assessment of this determines whether we study it carefully or not! Study may lead to reassessments but the assessment itself, right or wrong, is part and parcel of the dynamics of scientific progress! And so we should be liberal when assessing an idea; we should not say, this idea, this experiment, seems too revolutionary, too far out on a limb, so that we may safely ignore it. Rather, we may say, here is an idea or an item of information which I find intriguing. Is it intriguing? If yes, can we pursue study of it? If yes, can it be worthwhile? Is it worthwhile? The main thing I would like to suggest is that in our admitted ignorance we may care more for the young bright-eyed beginners, and try not to discourage them; this, I suggest, takes precedence over the concern with the welfare of science, especially since science is nowadays doing quite nicely anyway and its future depends on these bright-eyed beginners more than on its stuffed-shirt defenders.

The corollary to this is very opposed to Bacon's view, and it is due to Einstein. First, do not hoard information; ignore as much as possible, so as to get a perspective of the potentiality of items which look most promising, and then heavily concentrate on them for a while. Second, always be careful to allow the ignored items easy re-entry.

Einstein, the originator of this view, forcefully expressed it in his scientific autobiography. He had a great mathematical aptitude, but he feared that if he became a mathematician he might get grooved in one small sideline of mathematical research. In physics he tried to be ignorant of unnecessary detail. He crammed information for his final exams in physics; the result was that for about a year he lost all taste for physics; and all his life he tried to keep a clean broad view of the field as more significant than the possession of any piece of information. The result was most remarkable. Most historians of physics and writers of texts on Einstein's relativity were convinced that he was explaining Michelson's experiment of the eighties of the last century; he had no detailed knowledge of that experiment, however. But, as Gerald Holton has shown, Einstein was familiar with a study of the problem which that experiment created. The problem, now closely associated with Michelson's experiment, is what he knew about and centered upon. Likewise, Einstein invented in 1917 a theory which assumes induced emission—emission facilitated by the existence of radiation in the vicinity. Some years before, Wood and his colleagues experimentally observed such emission; Einstein was unaware of these experiments; nor did he need the information. Ironically, both Wood's empirical finding and Einstein's theory on induced emission were ignored, and all the textbooks of physics which I met as a physics student skipped them; this altered with the advent of lasers.

I have mentioned a number of problems which engaged students of radiation in the middle of the nineteenth century. The history of science has proven correct their conviction that these were important questions. And yet, the hero of radiation theory—and radiation theory doubtless has one chief hero—was Gustav Robert Kirchhoff, who cut through the thickets of all the apprehensions and the excess information and offered the central law of radiation named after him.

This is not to present Kirchhoff's idea of method as identical with that of Einstein. They could not be more different. But I am following Einstein's dictum: observe what scientists do rather than listen to what they say they do. Kirchhoff insisted that he had myriads of facts illustrating the fact that emission and absorption spectra are mirror images of each other. But his deliberations were a marvel of elimination of excess baggage, and mounting Einstein's methodology on Kirchhoff's researches makes them much easier to comprehend than they initially were.

3.7 The Place of Young in History

Young's great contribution to contemporary thinking is usually overlooked, as it comprises a change in philosophy, and one hardly noticed by philosophers proper. It is this: Young put human sensations in a clear physical context yet did not confuse sensations with their causes. This invites some elaboration.

The traditional philosophical attitude to experience was that of mistrust: until we learn to question our common way of seeing things, neither science nor philosophy is called for. But questioning is not enough: oriental philosophy recognizes the shortcomings of the world of appearances, declaring it sheer dreams, and leaves it at that, or adds to it a theory of reality and leaves it at that. Only western thought takes it as a task to explain the appearances. Somehow, it was taken for granted that thinkers had to engage themselves in the task of explaining. Too little is known about the ancient Babylonian astronomers who identified the evening star and the morning star—the first big discovery in astronomy—but this great scientific breakthrough did not suffice to establish science. It is not clear at what date it became clear to the ancient Greek thinkers that they were engaged in explanation, but sooner or later they were clear that this was their chosen task. Perhaps it was Parmenides. For, it was he who said, this cannot be done: everyday experience is full of error and eror cannot be explained by truth—everyday experience must be dismissed as mere dreams. The point was simple: *explanation is deductive, and error never follows from truth*. This is paradoxical, and the paradox still troubles philosophy to date.

The paradox was not resolved because it was more important to stress the unreliability of common experience than to find its role in science. It was hard enough for advocates of science to explain the claim of science that experiences, even the most basic ones, are not replicas. You might as well say that the tickle is in the feather, is Galileo's final word on the matter.

The matter did crystalize, however, into a doctrine, advocated by Demokritos in antiquity and repeated by Galileo and others: some sensations represent primary qualities, lile those of bulk and shapes, others, like colors and heat (and tickles) are secondary (and less). Why is the tickle less in the feather than its softness? This question was left fallow for generations.

The question was left fallow for generations because in the mean time a new school of philosophy evolved, to which Newton himself belonged and lent authority, and which declared reports of (carefully made) observations utterly reliable. To be precise, Newton never agreed with the certainty of observations as such: in his *Opticks* he discussed the question, what should be done about refuted observation reports, as he was intent on not discarding them entirely. Now the reports that he was concerned about were scientific, namely, generalizations; he meant that only the generalizations may be refuted as they may go too far, and so should be generalized more humbly. But singular observations are different. This is hard to maintain; first because they are not scientific, and second because his friend the celebrated philosopher John Locke had a refutation of this claim: put one hand in a hot dish of water, the other in a cold one, and then put them both in a lukewarm one; the one hand will feel cold, the other hot. Yet both of these thinkers had excuses for this strange fact.

Before Newton's advocacy of the reliance on sensations, the older dismissal of them had to be dismissed. This was provided by Robert Boyle, the only thinker whom Newton recognized as his senior. Boyle wrote a study of colors. He defended his work by declaring colors objective: I can dream the color of an object to be whatever I wish, but I cannot see it other than it is. Why, then is the tickle of a feather not as objective as its color? No answer.

This story has a moral to it: science can progress despite the presence of central and serious problems in its midst. But there always is some price to pay for unattended problems. And the example at hand is the difficulty with which the place of sensations became clear—as almost an afterthought. When Young discovered the existence of infra-red and of ultra-violet light he realized that here is a new attitude to sensations: sensations are events caused by some physical events, and the cause and effect should be clearly distinguished. This simple observation led him and Wollaston to their researches. Young's discovery of invisible light led Wollaston to the search for inaudible sound and to the discovery of variations in thresholds. It made them both search the differences between optics and physiological optics, beginning with the discovery of astigmatism and ending with the contributions to spectroscopy (Wollaston's discovery of the sodium yellow light and his suggestion that it is a mark between different colors, plus Fraunhoffer's refutation of this suggestion).

To see how far from obvious all this is, let me mention Mach's view. It is particularly significant its place is still not clear. Max Planck, the father of quantum mechanics, found it sufficiently important to criticize it, and many commentators found his attack quite unbecoming, since Mach was far from being hostile to science: he was a historian of science who ontributed significantly to both physics and physiological optics. Einstein declared indebtedness and admiration to him. Yet Planck was right to attack Mach's view as an impediment to research: He analyzed experience as sensations, after science had gone away from this and related sensations to the world as we best know it.

Not only the view of light as waves in the ether much more abstract; after all, the identification of light with ether waves and of sound with air waves were known long before Young; the difference he made was in his idea of waves as at times sensed and at times not. Thoug sensed qualities are caused by some kind of events, not all the events of this kind are sensed. This then shows that the clear distinction between a sensation and its cause may lead to the discovery of unsensed events. Experience is then a much more abstract matter than senesation. Many philosophers said that the theory of light as waves is a mere fiction, as a sophisticated way of reporting sensations, as a mere instrument; the claim that visible light is a part of a bigger whole reverses the order of things: not sensations accomodate the theory, but the theory accomodates the sensations as it describes sensed objects which abide in physical space: it was not experiences but what is presumably experienced that science examines. This helps resolve the paradox of explanation: explaining experiences we also learn to correct them.

Heat illustrates this better than ligt. High intensity cannot be sensed. This holds for light intensity tolerate, as Newton exhibited by looking at the sun until he nearly went blind. So meters of the intensities of light and heat were devised. So temperature was deemed to be a measure of heat, akin to the intensity of light which is a measure of brightness. This was progress: Robert Boyle deemed pepper hot and opium cold; The thermometer corrected this error. Still, the identification of heat with temperature is an error, and one still perpetrated by philosopehrs who follow Mach, such as Rudolf Carnap, and by most historians of science. And it is no easy matter to discard this identification.

This identification prevented for long the resolution of the dispute between the materialists (who said heat is matter of sorts) and the dynamicists (who said heat is motion). For, what causes the sensation of heat is not temperature but the rapid transfer of large quantities of heat. Carnot's theory solved the mystery, as I will narrate soon, as he introduced heat transport (entropy) into the picture. It is no accident that historians of science are caught in the dispute and try to take sides: they are unaware of the quiet revolution effected by Young in the year 1800.

The revolution spilled into other fields. Medicine was freed of the intuitive, home medicines, only after the intuitive perceptions of diseases were rejected, to make room for abstract ideas about diseases and their diagnoses. This was done not without opposition, such as the invention of homeopathy, the attempt to rectify rather than do away with ordinary intuitions. At the same time as medicine became more abstract biology took the final step with the theory of evolution, particularly in its Darwinian version of natural selection, which placed sense-organs within ecosystems. Yet the adjustment of perception theory to this change, begun by Young and ended with Darwin, took place only after World War II! This should give some feel for immensity of the discovery of invisible light.

Chapter 4

THE CHANGING SCENERY

4.1 The Wave Theory of Light

Light that falls on a material surface, incident light so-called, can pass through, or be absorbed or reflected; and normally, when a body opaque to visible light is irradiated with white light, then the colors it absorbs and those it reflects are complementary: complementary colors put together yield the color white. It seems that there is no option other than transparency, absorption and reflection, but there are. First of all we should notice that our concepts are vague, as things are more graded than thus far presented. Transparency is not quite a quality which is either present or absent; all bodies may be viewed as partially transparent and partially reflective of this or that wavelength of light at this or that temperature. There are other phenomena which may—but need not—be subsumed under the absorption or the reflection of light.

  For example, take dispersion. Dispersion of light is the phenomenon best illustrated with the dust particles dancing in a sunbeam in a relatively dark room. Dispersion is what happens on a windshield of a car moving eastward after dawn or westward at dusk when all of a sudden the windshield obscures the view when lit up by the rays of the rising or setting sun, unless it is unusually smooth and clean. For, when hit by light, it radiates; every scratch on it and every dust-particle on it begins to shine. This is not quite reflection since the scratches and the dust particles send light in all directions; they become secondary sources of light and look luminous indeed. Light beams are not visible in the ordinary sense of visibility. Ordinarily a body is visible either when it emits light or absorbs or reflects light. Air is usually invisible as it is transparent to visible light. It is visible as the red sunset on the horizon because at sunset light travels through more air than at midday and damp air transmits red better than other parts of the visible spectrum which it partly absorbs. (All this is correct in optics, not in physiological optics: a truly black spot—a speck of soot or of platinum black—is quite visible due to its not reflecting when placed among reflecting bodies! It is conspicuous in its absence, as they say of a conspicuously missing dignitary.) And so, light rays cannot be seen since they neither emit other light rays nor collide with other light rays. Hence, every time a light beam is seen, say passing through a hole in a cloud, what is seen is not reflection but dispersion. Dispersion then is a borderline case: caused by collisions of light with dust particles, it may be subsumed under reflection; but ordinarily reflected light is directed whereas dispersed light goes in all directions; as the obstacle, the dust particle or the scratch, looks luminous, it may count as a source of light. It is, we say, a *secondary* source of light. The only difference between dispersion and reflection is that the source of the dispersion is the secondary source of light, the item that looks as if it radiates.

Apart from dispersion there are other phenomena which are very useful for the study of spectroscopy. For example, refraction and diffraction. I am now in a peculiar position. For the purpose of the present study there is no difference between diffraction and refraction. The fact is that refraction looks different from diffraction in its historical setting and this setting ends before the beginning of our period, the nineteenth century. But I notice that the idea behind diffraction is so simple and easy, and comes so often into the background of the present study, that I would rather explain it. I shall be brief and untechnical, and I promise readers unfamiliar with it that they will be amply rewarded for but a slight effort.

Refraction is a simple phenomenon, naturally manifest in rainbows. Its simple explanation is that the refrangibility of rays of different colors is different. Diffraction, too, is rainbow-like, and the chief and impressive difference between them is this. When a light beam is passed through a prism, the different sections occupied by the different colors are not equal: the red is the smallest and the violet is the largest; in dispersion they are more proportionate. This fact is today deemed insignificant, but it played a crucial role in history as I shall soon narrate. Back to diffraction.

Diffraction involves interference but refraction does not. Interference appears on a smooth surface disturbed by more than one cause thus yielding wave patterns, as do the wakes of boats on a smooth lake. A peak and a valley of a wave can cancel each other. So, when two waves of the same length travel together, they may reinforce each other and become a wave of a higher peak or cancel each other and disappear. We say that they are in phase when they reinforce each other and out of phase when they cancel each other. All intermediate cases are possible, of course. When two sources of waves operate, since they alternate between the two extremes, lines of in phase and out of phase interference appear on the lake's surface. The same happens in diffraction except that there the diffraction grating, whether it is a grid of fine pieces of matter or a grid of lines scratched on a piece of glass, acts as the secondary source. The reason the grating acts as a secondary light source need not matter here. We may, if we wish, consider it a sort of dispersion. Since light of different colors disperses differently, the pattern of diffraction is a spectrum.

Newton knew of diffraction, of course, as it was already discovered by Grimaldi, an Italian mathematician of the previous generation. Since diffraction is a wave phenomenon, you may wonder, could Newton account for it? Of course he could, and he did.

Newton, we remember, denied that light is a wave phenomenon because, unlike sound, it hardly goes round walls. (In diffraction light does go around walls somewhat, especially in the diffraction of a light beam that just hits a knife. The question whether this invalidates Newton's argument and if so whether he was sufficiently honest is a topic that some historians like very much. Myself, I find it repugnant.) Sound, which is a wave phenomenon, goes round walls and other obstacles much more easily than light. Without going into detail, let me repeat and stress that, as Newton knew that light had many wave-characteristics, nevertheless, he mentioned its being so much different from sound as an argument against the view that it constitutes waves plain and simple. He agreed that light is associated some waves of the ether, and he even envisaged light-particles as eels waving their tails—the analogy is his—so that it would be a wavicle. This is very unsatisfactory as particles were then supposed to be the simple building blocks of the universe; as such they could not have moving parts. In retrospect Newton's theory of light seems to share so much with the wave theory that one may wonder what sort of a theory it is and why Newton needed his hybrid and insisted on it so. A.I. Sabra wrote a fascinating book on this question. It turns out that what looks to us a most obvious wave-like phenomenon was, according to Newton, quite to the contrary, the clearest evidence that the wave theory will not do.

Newton preferred his corpuscular explanation of the fact that the grating is a secondary source to the wave theoretical explanation of his opponent, Huygens, since the wave theory could not account, you should remember, for the difference between light and sound. Light corpuscles collide with the grating, he said, and at times ricochet from it.

This theory led him to a strange explanation of the difference, which I have mentioned before, between the spread of colors in diffraction (where it is even) and in refrangibility (where it is not). According to his theory both were ricochets of sorts. Therefore he took more seriously than we would today the fact that in refraction but not in diffraction the spread of light at the violet end of the spectrum is broader than at the red end. He tried to find a deep theory behind a formula relating the diverse range of the refrangibility of colors which, he proposed, made light akin to sound, less as the wave theory has it and more as music theory would, that is, as a kind of harmony. (The reference to music should not be surprising. Kepler wrote about music, and so did Descartes, as they viewed music and science as very close to each other.)

I have described in the first chapter of this book the phenomenon known as Newton's rings (although he was not the first to report them—Robert Boyle and Robert Hooke did so before—and I do not know who discovered them first). The rings are multicolored and appear underneath a semi-sphere of glass, or underneath a water-filled round bowl, when it is lit from a point right above. It is easy today to comprehend this phenomenon as a refraction: just think of a cone as a circular prism, and then think of half a ball as a prism. Yet it is Young's discovery of this that has led to the revolution in optics as I shall soon explain; Newton did not know this and so he puzzled over the difference between different spectra. He faced two difficult questions:

(a) what causes interference if not waves?
(b) what causes the different colors to spread differently in refraction?

These questions are not as Newton (mis)understood them. As for question (a), diffraction is a wave phenomenon, and even Newton accepted this, but he attempted to correlate the wave aspect of light with its being corpuscular. As for question (b), it is, simply, why do different colors refract differently? In other words, why do different colors have different indices of refraction for the same kind of matter?

Young, we remember, was drawn to study interference when working on acoustics. The link between optics and acoustics which guided him is the theory of harmony. This part of Newton's theory is not mentioned in common histories of science. They hide the strange fact about Newton's study of colors, his attempt, on the assumption that there are seven colors, to calculate their spread on the (prism) spectrum in accord with some general mathematical law. This, we remember, means in modern terms that each color is between two indices of refrangibility, and so we have a series of numbers representing colors which we may put into a formula. Such a formula, relating to sound (Pythagoras) or to planetary distances (Pythagoras and Kepler), is called a harmony. Young moved from acoustic harmony to optic harmony and from it to the wave theory of light and we should see his way of reasoning when we are familiar with the situation in which he found himself. This is not to say that he was sufficiently successful; indeed, he was not. But his success in explaining Newton's rings by the application of Newton's own theory of diffraction was for him a success that invited more attempts in the same direction; he tried to explain diffraction and refraction as one and the same phenomenon, quite contrary to Newton, but still within the Newtonian corpuscular theory. But once refraction and diffraction were put on equal footing, Newton's puzzlement about their different spreads was easily solved without recourse to his theory of harmony. (It is no accident that Fresnel, too, broke away from Newton's corpuscular theory by studying diffraction though his way to heresy was different and came form a metaphysical idea: all ethers are identical.) Following this line of investigation, Young found, to put it in modern parlance, that refraction is not a step function but a continuous one, namely, not a magnitude for each color but a magnitude which varies gradually with the change of wavelength. This, we have seen, makes the corpuscular theory of light lose its charm as this forced the theory to assume the presence of infinitely many kinds of light particles, contrary to the principle of simplicity.

To conclude, the harmony theory was Newton's way of accounting for the similarity he found between light and sound (whatever this is); it is thus not surprising that Young's optical discoveries expelled the idea of color harmony, as they did away with the seven colors or the optical scale. Or should I attribute this development to both Young and Wollaston? After Young was practically ostracized, Wollaston came to his aid. They were then very close collaborators (almost all the way) in both optics and physiological optics—in both his theory of light and of color vision—and together they totally changed the scenery.

Harmony did return to spectroscopy, but only about a century later with the work of Balmer and more so with the atomic theory of Niels Bohr. (Einstein, in particular, was deeply impressed with Bohr's harmonies.) But this is the end of our journey and it has nothing to do with Newton's harmony of colors. Helmholtz suggested (in his essay on Goethe's theory of colors) a different theory about the harmony of colors based on Young's theory of color vision, but it was a fiasco (like Goethe's theory which he found embarrassing).

A paragraph on Young as a collaborator. Young's theory did not get off the ground, despite some magnificent success, because he could not answer Newton's objection: unlike sound, light does not go round walls that easy! And the way this difficulty was countered is the story of the second stage, due to Fresnel, a younger and distant colleague. Some historians of science stress this fact but do not explain it though the explanation is simple: Young was in error as light waves are very different from sound waves. Others overlook the difference between the two stages, since Young was a brave pioneer (of course he was, and concealment does him no honor). The first stage was the collaboration between Young and Wollaston, a senior colleague who could take care of himself. As to the second stage, Young was ambivalent about Fresnel, but finally took a positive step and translated the only paper of his to appear in English that time. Fresnel did complain in a private letter to him that his share was not fully and justly appreciated. I cannot judge this, but I did find the letter very moving. It is obvious that to acknowledge Fresnel's contribution is to contrast it with that of Young, and this policy was not acceptable at that time and is still not popular. It was done only recently by Geoffrey Cantor whose study of theories of light simply presents the two theories as different. To prevent suspicion that I am too generous to Young in concealing his lack of criticism, I should add that the very suggestion, which was so different that he could scarcely acknowledge the difference, the suggestion that light waves are transversal, came from Young himself who wrote to Fresnel in 1817 about it as a possible explanation for polarization (to which we come soon). This is the difference between a metaphysical idea, one that serves as an intellectual framework, and a scientific idea: the one is easy to comprehend and serves as a challenge; the other should be worked out in detail or else its value is unclear. Fresnel was very able at calculating and Young was a great inspiration.

Let us return to a point made by I.B. Cohen about the light which science can throw on its past history. We should not evaluate old of theories by their accord or discord with current ones, as such comparisons do not throw light on reasoning behind any idea. But such comparisons may tell us about the difficulties which early thinkers had met. Consider the question, what exactly makes light partly wave-like and partly corpuscle-like? Einstein's partial return to Newton's theory of light, says Cohen, should not make us rewrite its history, but it can and should make us raise our appreciation of it.

4.2 More About Waves

Polarization is the next quality of light on our list, next to its absorption, transmission, and reflection, as well as its dispersion, refraction and diffraction. Polarization is now generally known due to the popularity of polarized sunglasses. Two polarized lenses put together may be transparent or opaque, depending on their relative position: they transmit light when placed "in parallel", but not when placed "in perpendicular". When "in parallel", they transmit light maximally, and this is exactly half of the incident intensity; when one is fixed and the other is rotated they soon block all light. The polarization of light waves was discovered in 1808 by Malus who found that light has this strange property when reflected from any reflector at some given angle. The very attempt to explain this phenomenon, even to present it easily and correctly enough, requires a totally new theory of light-waves. This was the next stage.

In light of the polarization of light Fresnel suggested that this phenomenon illustrated an essential difference between sound and light. (So Newton had a point after all!) Sound waves are longitudinal: the particles of air move to and fro in the direction of the movement of the sound wave (along the line drawn between your mouth and my ear). Light waves are transversal: the ether-particles move to and fro in the direction perpendicular to the movement of the light-wave (just as water waves on the surface of a pond though in three dimensions). In three dimensions we have as many directions perpendicular to the direction of the wave as desired. It is sufficient, however, to consider only two directions (coordinates), preferably perpendicular to each other, and view the transversal wave as two waves, each moving along one of these two directions (coordinates). Ordinary light waves, Fresnel suggested, oscillate in both transversal directions (i.e. along both coordinates), but polarized waves oscillate only in one plane of polarization. When a light wave moves through a polarizer, energy along one direction is absorbed and that along the perpendicular one is let through. When another polarizer is placed perpendicularly to the first one, all light is absorbed. (This thrilling case is even more involved. If you have a good geometrical intuition, try this: usually the two polarized wave trains perpendicular to each other are in phase; when entirely out of phase, the result is circular polarization.)

The importance of the idea that light waves are transversal is historically very great. Gases vibrate only longitudinally, but elastic solids vibrate both longitudinally and transversely. These two vibrations propagate with different speeds calculable on the basis of the laws of elasticity. Newton thought that the ether was a gas, or rather, he did not distinguish well enough between a gas and a fluid; according to Fresnel it is solid. Elasticity became an important field of study. The search for longitudinal waves of ether went on for a century (at first Röntgen thought he had found them when he discovered x-rays). Sooner or later, however, with accumulating knowledge of both elasticity and the elasticity of the ether, the ether was given up altogether as too problematic. Fortunately, the problem of radiation that it gave rise to remained central. But about all this later. Meanwhile, let us examine light on the assumption that it is undulate, i.e., a wave phenomenon.

Waves have velocities, lengths, frequencies, and amplitudes. "Frequency" is a very common concept. High frequencies relate to more cycles or repetitions per time unit, low frequencies to less. Every quantity which is multiplied by repetition will be the product of the quantity contributed each time multiplied by the number of occurrences of the repetitions of that contribution. If a pendulum oscillates 30 times per minute or $\frac{1}{2}$ times per second, and if it travels 1 foot in each direction, or 2 feet per oscillation, then it covers 60 feet every minute or 1 foot every second; its average speed is 1 foot per second. Now a wave which is as regular as a pendulum is called monochromatic (of one color as the color of light is determined by its frequency); its wavelength—the distances between its crests—is constant, and its speed is plainly the product of its length and its frequency. Put in a formula, let c stand for the velocity of light, and let the Greek letters λ (read, *lambda*) and ν (read, *nu*) stand for wavelength and frequency respectively; then,

$$c = \lambda \times \nu.$$

(As color depends on frequency, and as it is possible to slow down light, e.g., by passing it through water, it is possible to change wavelengths without changing color.) Real waves are never monochromatic; until lasers were discovered around the middle of the twentieth century, monochromaticity was achieved only very approximately (by the use of filters, which always "smear" what they funnel.)

Monochromatic waves do not exist. This might have stopped all research in optics as soon as the wave theory was established (after 1821). (Young's classic work is of 1800, Malus discovered polarization in 1811, and Fresnel started his campaign a few years later. His work on transversal waves is of 1817; victory came in no time—officially in 1818.) Fortunately, work on *approximately* monochromatic waves went on in a lively manner on dispersion phenomena (Fresnel), and on indices of refraction (Fraunhofer) and on the velocity of light in different media (Fizeau).

Meanwhile Fourier had overcome the difficulty in his 1830 treatise on heat transport. Why heat transport became important will soon become clear. Yet his result proved important in a very general way as it concerns waves in general, any waves whatsoever. This is Fourier analysis: in an attempt to describe a wave phenomenon, however complex, it can be viewed as roughly a simple sum of different monochromatic waves; the more monochromatic waves one is allowed to add, the better the representation becomes. Here it is easily noticeable that the wave's amplitude comes into play.

An amplitude is the size of the crest of a wave. Obviously, when one wave is mounted on another wave of a much greater amplitude, then the added wave may be ignored without much distortion. Fourier suggested that when a wave formation is described, its components which possess relatively high amplitudes should be taken into account first, and then its components which possess ever smaller amplitudes may be admitted into the picture so as to increase precision. Thus, it is possible to create a series of pictures of a given complex wave by considering more and more components, and this series of pictures rapidly converges towards the real wave thus pictured. In principle, however, a genuine wave in a genuinely continuous (non-atomic) medium may be viewed as a mixture of infinitely many monochromatic waves!

What this amounts to is very interesting. Newton, who had envisaged white light as composed of seven different kinds of light each possessing its own degree of refraction, suggested that instruments may separate the light particles of the different kinds because they are mixed in the white light even before their separation. Newton's celebrated experiment with the separation and mixture of light illustrates this clearly. There are numerous experiments in which light is separated and mixed, all in the effort to present phenomena that cannot be explained except on the supposition that white light is a mix. (The process inevitably reduces the overall intensity of the light, and this led the great poet Johann Wolfgang Goethe to rebel against Newton: gray is not white, he thundered, to the embarrassment of his admirers, you remember.)

What exactly has happened here? By Newton's theory, white light is nothing but the mixture of seven colors; he separated them (with a prism) and then mixed them again (with a lens). Whiteness is in the eye of the beholder. Not so according to the wave theorist whose view can be better illustrated with an example presented by the great twentieth century physicist Sir Arthur Stanley Eddington. We can say that the material of a statue was in the stone from which it was chiselled, but not that the statue was hidden in the stone: the statue is matter plus form, form imposed on matter. So with light. Light is the energy of vibrations: monochromaticity is the simplest vibration, a sine wave so-called; color, chromaticity, is form, a recurrent pattern; a beam of white light is undifferentiated or unformed energy. Hence, a grating or a prism imposes form on energy; the form was no more in the original beam than the statue in the stone. And so in Newton's experiment the oncoming white beam is unformed, the first prism imposes order on it, and the second erases the order and brings back the initial chaos. The first prism creates or shapes colors the way an artist creates a statue or forms a piece of clay.

Assuming that light is energy, what is that energy? It is a wave energy, and the energy of all waves, whether of sound or of light, has at least one shared characteristic. It is this. Waves carry kinetic energy. According to classical mechanics, the energy of a wave, any wave, is proportional to the square of the amplitude of that wave. (The bigger the crest the more energetic it is, regardless of whether its recurrence is regular or not.) The energy of white light (or of white sound for that matter—the sound of a waterfall, for example) is a combination of energies of many other kinds of light (or sound), of monochromatic waves. But this is sheer abstraction. When white light hits an obstacle, a prism or a grating, the energy of the original disturbance is funnelled in various directions and in each direction in a fixed frequency. Sound does not exhibit such beautiful patterns as it easily goes around walls, so there is no prism good enough for it, though there are acoustic lenses, filters, absorbers and reflectors that exhibit some similarity between sound and light (too small to impress Newton, you should remember).

Fourier's analysis enables one to present light as a combination (in the abstract) of (possibly many) monochromatic rays of different wavelengths (and hence frequencies) and amplitudes, all moving in concentric straight lines except when the medium changes in a way which alters both speed and direction in a simply correlated manner.

4.3 Light Waves and Matter

Much of the wave theory of light has little immediate correlation to radiation theory. The speed of light never entered considerations pertaining to radiation. The speed of light in different media was a problem relating, of course, to theories of matter and its interaction with light. Electromagnetic theory has explained some of these phenomena, and the others were explained much later by the theory of relativity. No doubt, the explanations offered by electromagnetic theory were not entirely satisfying; that theory accepted as unexplained empirical fact, especially the magnetic and electric properties of matter. Anyway, it is impossible to explain varying speeds of light without taking account of the fact that a medium is transparent to one color and opaque to another.

Here we see what can happen when we are unable to meet a question, such as, what causes changes in the speed of light? We may switch to a bigger question, namely, how do light and matter interact?

Leaving the speed of light, we leave along with it the direction of light since the direction changes in the process of refraction which takes place during transition of light from one medium to another as a result of a change of speed. (Replace a wavefront with a row of runners and suppose that the flank of the row gets off the highway into rough terrain and are consequently slowed down, and you will see at once that as a result of the uneven speed the row will change its direction.) Leaving refraction, we have still diffraction—the ricochet phenomenon. Strangely, diffraction bears no relation to the material causing it. In the case of refraction the material of a prism and its surrounding determine the index of refraction of the prism in that surrounding and thus the spread of the spectrum. Not so in diffraction. Light passing near the edge of a knife or through a grid—whether of fine wires or of fine lines scratched on a piece of glass (as found by Fraunhofer)—diffracts independently of the material from which the diffractor is made. All that matters for diffraction is that the diffractor is an obstacle. Why is it an obstacle? Why is a scratch on glass an obstacle? These are most difficult questions. But given an obstacle, the wave theory of light tells us that it diffracts light.

The same goes for dispersion which is explained like diffraction; dust-particles and their like are small obstacles.

Polarization is the fact that when light passes through certain crystals, or is reflected at certain angles, the waves which previously oscillated in all planes perpendicular to the direction of their propagation are now reduced in energy by half (through absorption), and cut down to oscillate in only one such direction. The study of this phenomenon is very interesting, but going further into its mechanism belongs (mostly) more to quantum theory proper than to radiation theory. I shall avoid this topic as much as possible.

There is one more question: can the interaction of light with matter alter its frequency? The wave theory of light says no. The facts of experience seemed to say yes. Let me briefly explain.

The first fact is the one known as fluorescence, discovered by Herschel and explained by Stokes. Today it is easiest to perceive the phenomenon by noticing that teeth and nails shine when they are exposed to sunlamp-light in a dark room. The room is dark because the light that the lamp emits is ultra-violet which is to say invisible. The teeth and nails shine because they render the lamp's light visible. Generally, and this is Stokes' law, fluorescence lengthens the wavelength of the light it involves. The wave theory permits a body to interact with a wave only if it has the same frequency as the body in question; all interaction between waves and elastic bodies is thus a resonance phenomenon. Hence, Stokes concluded, fluorescence is more complex a phenomenon: matter absorbs, pauses, and then emits; hence, the emitted and the absorbed light differ; hence, their frequencies may differ too.

Before criticizing Stokes, let me observe that he hit upon a pattern that has become the central one for emission and absorption. An atom which absorbs energy gets excited; it can get excited in different ways by absorbing light or thermal energy or chemical energy or any other kind of energy. Only excited atoms radiate since radiation is a loss of energy (we ignore nuclear energy here).

Back to Stokes' theory of fluorescence. It looks as if Stokes reached a true conclusion from a false premise, always logically possible. It is even true (for all we know) that fluorescence is a process of absorption and re-emission. And yet, Stokes was mistaken. Contrary to the classical wave theory, it is possible for matter to alter the frequency of light without first absorbing it—while colliding with it. This was predicted by Einstein in 1916 and observed by Arthur H. Compton in 1922. Indeed, we call the diminution of the frequency of light due to collision with matter the Compton effect though we go on calling the diminution which is caused by absorption and re-emission (or a combination of these and other events) fluorescence. And Compton's effect was declared the empirical *coup de grace* to the classical wave theory, even though it, too, later did undergo some important changes.

We are not done with Stokes. He noticed something forbidden in fluorescence and pushed the forbidden fact out of sight. The classical wave theory says that a wave does not change its frequency when it meets matter—whatever else may happen there. Well, said Stokes, we do not know much about absorption and emission, so let us say that matter absorbs light and then emits light of smaller frequency. This is, indeed, Stokes' law. But the classical theory of elasticity was further developed and it turned out that absorption and emission are resonance phenomena. A piece of matter absorbs, emits, and resonates on one and the same frequency; it is the body's natural frequency and its overtones or harmonics. Because matter of the same chemical constitution emits the same frequencies come what may, we know it is the atoms which emit and not the whole bodies.

For example, a tuning fork or a string on a piano has its natural frequency, i.e., the frequency which it sounds when it is excited, and it can absorb sound only of the frequencies it can emit, but these frequencies depend on the conditions of the sounding bodies, so that strings made of the same material can produce different sounds (or else we would not have our familiar string instruments). But there is no way to produce different colors from a piece of metal, as the radiation that it emits comes from its atoms! Light is electromagnetic and so it interacts with the atoms because atoms are electrically charged. Ampére discovered that all atoms are electrically charged (1820). How? Light interacts with the electrons which atoms contain and which also vibrate. And by the classical theory of electromagnetism, Maxwell's theory, light is the elastic waves of the ether. And so Stokes' law does not agree with Maxwell's theory either. But this is not to blame Stokes.

For it was about half a century prior to the rise of quantum theory that there developed an accumulated store of information about absorption and emission where matter absorbed light of one color and emitted light of a different color in exchange. This is not a fair exchange: something gets changed which ought to stay put. And yet Prevost's followers went on doggedly ignoring all this mass of information: they could not figure out why or how all this happened and hoped one day to hit upon a nice conservative solution to the riddle. When the solution came at last in 1913 with Bohr's revolutionary model of the atom, classical physics had already been superseded; quantum theory came to stay.

All this is later history and I am anticipating myself. Let me only mention, then, a few kinds of phenomena that were totally outside theoretical physics until the advent of quantum theory and then proceed with the story.

In fluorescence there is a delay between absorption and emission. In some cases a long time, hours or days, passes between absorption and emission. These special cases of fluorescence are called phosphorescence.

There is a broad range of interaction between matter and light, of absorption and emission to be more specific. Absorption need not be heating though it must be an increase of some form of energy. Absorption may lead to chemical changes (photochemistry) and chemical changes may lead to emission (luminescence). Absorption may lead to electric changes (photoelectricity) and electric changes may lead to emission (scintillation). Some of these phenomena were well-known and, since Priestley's classical studies of photochemistry they remained subject to intense studies; these led to the invention of photography in 1839 or thereabouts.

As these transitions embrace energy exchange, and as the known concept of energy was insufficient to explain them, a better idea of energy was invited: the conditions under which any one of these transitions takes place should be specified and explained. A particular puzzlement is presented by gravitational energy: it can be converted only to kinetic energy; it is not known how to convert gravitational energy to, say, electric energy. (In hydro-electric plants gravity is first turned into motion and then to electricity, not directly to electricity.) I do not know how you feel about all this. If you find it confusing, I cannot blame you; indeed, I share your complaint. So I promise to try and sort out things a bit more slowly from now on. I hope, also, that you find all this sufficiently intriguing; and I hope that you have noted the idea that each form of energy can in principle convert to any other form of energy, but that we do not know how to illustrate this principle in all cases. We may keep trying, though. This idea was conjured by Hans Christian Ørsted, the father of electromagnetism, and Michael Faraday, the inventor of the electric motor and the dynamo and the father of field theory. I shall discuss field theory after sketching the new views concerning heat. Classical quantum theory is presented with scarcely a reference to thermodynamics; radiation theory has thermodynamics at its very heart.

4.4 Heat as Energy

Two traditional theories of heat have competed since the seventeenth century. The one theory is materialism (the word was invented in the seventeenth century by Robert Boyle); heat is a kind of matter. The other is the kinetic theory: heat is the motion of the parts of hot bodies (like the motion of parts of boiling water). They were both superseded in the middle of the last century.

Taking for granted that it is condescension and thus disrespect to forgive our predecessors their intelligent mistakes, let me say that, though materialism is undoubtedly false, the kinetic theory of heat may be true or false depending on what exactly motion is. And almost all answers to this question, except perhaps the most recent ones, are known to be false too. This fact is noted only by the more advanced historians of science. Most historians of science accept the authority of science and so they take it for granted that heat is kinetic energy; they declare that kinetic energy is motion. Not so: it depends on motion, as Leibniz has shown already in the seventeenth century. (Leibniz called kinetic energy *vis viva*, namely, living force. He applied it to Galileo's theory of the pendulum to show that when it disappears it may be retrieved.) The famous equation for energy,

for *closed systems, kinetic energy + potential energy = constant*

is eighteenth century: Lagrange and Laplace proved it within Newtonian mechanics. (Incidentally, this mathematical discovery reformed Newton's view; he disagreed with Leibniz and denied that energy is conserved.)

The (erroneous) view that today's science textbook considers heat as energy has led some historians of science to view the materialists as the losing party, as the thinkers who backed the wrong horse; so they dismiss them as prejudiced. More daring historians defend materialism since some materialists have made significant advances in the field; They allege that materialism was true then, as if truth is not timeless. Very few historians of science view both materialism and kineticism as false and interesting—as valuable errors.

The truth of the matter is that the historians of physics who try hard to follow the up-to-date physics textbook often fail. First, they understand the physics textbook in a strange way. Whereas the quantity of heat is the quantity of kinetic energy, according to the physics textbook, the degree of heat, the temperature, which is what we usually mean by heat, is neither a kind of matter nor energy, kinetic or otherwise; ***heat is the concentration of kinetic energy.*** The traditional descriptions of heat have been thoroughly replaced by descriptions of temperatures and quantities of heat, of specific heats, of entropy and of other quantities associated with heat seldom mentioned outside a very specialized literature. Quantities of heat are more-or-less quantities of kinetic energies, not at all temperatures, nor the other quantities associated with them. (This was explained already in Chapter 1 above.) The second error of almost all historians who speak of the matter is their oversight of the contributions of disputes between rival parties to the advancement of science.

The historians of science who take the materialist theory of heat as a basis of some serious historical studies, mainly Thomas Kuhn and his followers, deny that the materialist theory is false; his doctrine of incommensurability, so-called, patronizingly declares competing theories non-competitors, thus willfully distorting history. As my quotations below from Carnot amply illustrate, the development of ideas was dialectical: the dialogue between researchers is what keeps them progressing the way they do. The progress at hand is the development of the theory of heat into theories of the various quantities associated with heat phenomena as mentioned above. For example, many historians of science report that around 1800 Sir Humphry Davy proved the theory that heat is motion by experiment: he created heat by friction, and that Count Rumford did so even earlier. I do not understand all that happened in the history of theories of heat, but I find this statement quite clearly erroneous though it is often repeated. There was no need to wait for Rumford or Davy to find that friction causes heat. This, you remember, was well-known to the materialists, not to mention their ancestors who used to rub sticks together to make fire, and rub hands and stamp feet standing watch on cold nights.

Let me make a general point here concerning the history of science. It is very likely that each of the extant competing views has a striking illustration supporting it. Just as in Rumford's case, the kinetic theory is illustrated strikingly by the production of heat in friction, so in Lavoisier's case the materialist theory is illustrated strikingly by the production of heat in combustion. Both schools struggled with the known fact that the production of heat is proportional to investment. But most historians of science follow Bacon and share his contempt for controversy. They decide that one side in the controversy was always in the right since the truth is absolute and so timeless. These historians tend to ignore the striking evidence in favor of the party they think poorly of. They likewise fail to ask, how striking the evidence really is? If friction causes heat which is therefore motion, then, since friction also causes electricity, electricity likewise should be motion. (Indeed, this was an opinion voiced at the end of the eighteenth century, and it was one that Faraday probably held all his life.) But it is not; at least the fact that friction electrifies does not convince these historians that electricity is motion. Hence, their argument is invalid: the evidence is not that striking. Other historians of science follow Pierre Duhem's idea that the truth is relative, usually in the up-to-date version due to Kuhn. They allow themselves to praise a party which has lost in a scientific dispute, but then they strangely deny (or ignore) the obvious fact that the party which they praise has lost. In a clear sense, admittedly, the calorists lost in a way that the kineticists did not: unlike the concepts of motion and of energy, the concept of the caloric (and of the ether) is outdated. But in a sense no less clear, both parties have lost the dispute, though their heirs are still continuing to participate in debates that are as often rejuvenated as theories come and go. Kuhn's historiography is an advance in that it permits the presence of contending schools of thought, but only as a preliminary stage. He regrettably denies that science is constantly in a preliminary stage, and that disputes between contending scientific schools promote progress.

Why do historians of science find it so hard to praise wise people who have lost in a debate? No other respectable historians find this difficulty concerning any non-scientific dispute. Only historians and philosophers of science, when they discuss science (in the abstract or in history) find it impossible to praise the losing party (in flagrant violation of the historical fact that sooner or later all parties to every scientific dispute except for the most recent ones have lost).

It is much more interesting, generally, to pay little attention to what one party says about evidence in its favor, striking or not, and to ask, rather, what did they say about the evidence which the opposing party has adduced against their views? In other words, how is criticism faced? How did Rumford explain the fact that combustion heats? How did Rumford and Davy account for temperatures rather than for quantities of heat?

(This is central to the philosophy of Karl Popper: what counts in science is not evidence in favor of a theory but the readiness to face criticism and even to be self-critical.)

Rumford's and Davy's criticism was answered by one important materialist or calorist (Lavoisier, you do remember, called the matter of heat caloric), Sadi Carnot, who defended brilliantly the materialist conclusion that in experiments which produce heat, whether by friction or by any other means, caloric is transferred from the environment to the heated body (since matter cannot be created, as you remember). He did so by considering seriously the idea that friction heats; it is not only friction, he noticed, but any kind of work; his new discovery was that in order to raise temperature work is required. In return heat produces work when it flows back from the heated body to its colder environment. Planck called it the second law of thermodynamics as, having called the law of conservation of energy the first law of thermodynamics.

It is hard to judge how much of all this was known before Carnot. In a sense it was common knowledge, though never clearly crystallized, but Carnot added to it a point for which he won (posthumous) fame. By a celebrated thought experiment he deduced from this crystallized knowledge the conclusion that the higher the temperature difference between source and sink of heat (between the inside of a heat engine and its exhaust), the more efficient the work due to the heat flow. The kineticists were stymied.

Carnot's work was revolutionary in many respects. It introduced thought experiments as a powerful tool in a new manner. (Galileo had used an important and famous thought experiment two centuries earlier. But Galileo's thought experiment was not deemed an integral part of science as it was a critique of Aristotle's pre-scientific view; it could be deemed redundant.) Carnot's thought experiment showed that the debate on the nature of heat had been defective; to become more serious it must take into account entropy, the work performed by heat transport (the word "entropy" is Greek for transport).

Carnot's thought experiment is very important for our case study both because it changed the theory of heat in a most significant manner and because it rendered the concept of thought experiment a common fixture. Kirchhoff and his followers used it repeatedly. His idea was that there is a limit to the efficiency of transforming heat into work, contrary to Rumford's view that transforming heat and work is not limited. Without going into the details of the calculation, the general idea can easily be presented: heat can be transformed into work only up to a limit even when we ignore all loss due to the escape of heat into the environment and due to the fact that heat engines heat the environment by transporting to it heat from a hot reservoir. The limit of efficiency then is due to the fact that as the heat engine works, its reservoir cools down, and when its temperature falls to that of the environment, it cannot be exploited any more; moreover, some of the work has to be reinvested to return the engine to its initial state.

Carnot was not so pleased as one might think. True researchers are more interested in the truth than in an easy victory (or a difficult one for that matter) over opponents; they are very sensitive to difficulties. Carnot died very young. His notes were recently published, and there he expressed worries over Davy's experiments which produce heat by friction. So let me go over the argument a bit more slowly.

What the materialists must acknowledge is that the quantity of heat of any closed system must be finite since they view heat as a substance and substance is unalterable. When materialists saw heat vanish, say while melting ice, they said that it is not destroyed but hidden. Rumford and Davy said that since as much heat as desired can be produced by friction within a closed system, heat is not a substance. (Rumford argued so explicitly: if heat is a substance, then it cannot be created.)

The argument does not hold for reasons well-known to all parties then and the law of conservation of energy only makes it more obvious: the system in which heat is produced is either open or closed; if it is closed, then the experiment is limited so that it cannot produce unlimited quantities of heat; if it is open, then it gets heated by transporting something or other, be it caloric, motion, energy, or anything else.

Yet mistaken as Rumford and Davy were, they revealed the poverty of the theory of heat of their day; thus their work set Carnot thinking.

The more sophisticated theorists who worked after Carnot made the two moves that he was the first to try out; one was a combination of both old theories, namely, the idea that heat is the motion of caloric; another was that heat is the elastic force of caloric (rather than that heat is caloric). Both took cognizance of Prevost's identification of heat with radiant heat. But this was not easy and finally it led to the collapse of classical physics and the rise of quantum theory since, though Prevost was in error and heat is not radiant heat, radiant heat is heat, and incorporating this fact into thermodynamics was the task of radiation

Chapter 5

KIRCHHOFF'S LAW

5.1 Spectral Analysis

Spectral analysis is the best and easiest way to discern the presence of even elusive traces of any chemical element; spectra are like fingerprints. This was suggested by Fraunhofer already—in 1817 and in 1820. It got first established forty years later by Bunsen and Kirchhoff, and at once Bunsen used it to discover new elements.

The reason the fact was not easily accepted is that it is overwhelming; it would have certainly been more easily accepted had it not been so incredibly powerful. You remember Fraunhofer's yellow sodium line, I hope, the sodium D doublet. It turns out to be the spectroscopist's worst weed—like dandelion; it is ubiquitous either because it is not peculiar to sodium or because sodium is ubiquitous. But how ubiquitous can sodium be? Surely, even if there is sodium everywhere on earth, somewhere on earth it must be extremely rare; would you suppose that even such vanishing quantities should show up in spectra with a strongly marked yellow line? At first this was considered a most fantastic claim. But then Swan showed it to be the case. Then Bunsen found the spectral lines of new alkaline metals which are ubiquitous too, but in small quantities and constantly overlooked because they look so much like other alkaline metals. Soon spectroscopy enabled people to discover other trace elements, especially rare earths.

Already in 1835 Charles Wheatstone showed that electric sparks passing between metallic poles exhibit the spectral lines of the metals from which these poles are made. Ångström showed in 1853 that sparks also exhibit the spectral lines of the gases between the poles. Though under high pressures gases emit continuous spectra (this is Newton's experiment: a glass phial filled with smoke glows when heated), under low pressure they emit discrete spectra. Both the sodium and mercury arc lights and the neon and argon lights (which add so much to the inelegant quality of the modern city scene) are colored in this very special way: they emit discrete spectra due to mere traces of these gases in the tubes and in the arc spaces that are practically vacuous. They also consume very little electricity; the electrons that rush through the vacuum from time to time bump into the neon or argon atoms and transmit to them some of their kinetic energy which they soon give off as light of very specific frequencies.

The glowing tubes were invented *circa* 1860 at the end of a long, arduous road trodden by many researchers in quest for the best ways to produce sparks and spectral lines of diverse chemicals, solid, fluid and gaseous. The chief reason for this is that it is so easy to detect impurities spectroscopically; the sought-after lines are chiefly of impurities, of elements which enter the picture all by themselves.

On top of the spectra emitted by heated elements, there are absorption spectra. And the major victory of spectroscopy is the idea that absorption spectra are negatives of emission spectra. Thus we can identify those elements in a fluid by passing light through it and looking at it through a spectroscope. But, of course, matters can be done more delicately; concentrations of a given element in an absorbent material can be correlated with the relative strength of its absorbing spectra.

Sunlight is full of dark spectral lines or absorption spectra—this was the beginning of our story, you remember. It was Wollaston who discovered the first solar dark lines in 1802 and Fraunhofer who discovered the bright ones in 1817 and identified the dark and bright sodium D doublet in both. Historians of science tend to report or assume that their readers know that the absorption spectra are created by the sun's atmosphere, the sun's corona (meaning crown) which radiates light but which also absorbs light and thereby causes absorption spectra. This sounds as if the physics of the solar spectra is no longer problematic; it still is. Worse, the information just mentioned tends to obscure the way things looked before this information was known and deprives the reader of the thrill of the discovery as well as the comprehension of the process of thinking that went into it. So let us view the world without too much (ill-digested) up-to-date information.

Is there sodium in the sun then? It is easy to rush to that conclusion and say, yes. But, as sodium is ubiquitous on earth, perhaps the solar sodium dark line comes not from the sun but from the atmosphere of our own planet. I shall not go into all this; I mention this to you as a warning, as a token example of exactly how all histories of science, including this one of course, are gross and unfair distortions of historical records by their very selectivity, in particular, by the omission of one important stage in every development. Any discovery can more easily be explained away as a result of an uninteresting factor rather than something genuinely new and exciting; the initial suspicion is that a report is less exciting than it sounds. To repeat the observation of Mario Bunge, taking cognizance of the less exciting possibilities first, at least the more obvious among them, and eliminating them, may raise the level of excitement. Historians of science, however, need not bother themselves with this. For example, the contribution of the earth's atmosphere to all our astronomical observations—always a troublesome contribution—can now be fairly easily studied by comparing some observations made on earth with some made on an artificial satellite or on the moon. Such lovely means were hardly conceivable in the last century. Given the solar dark lines, how would you go about answering the question, which of them is truly solar rather than terrestrial? Still harder, how much is each truly solar rather than terrestrial? If you are very smart, you will say, by comparing solar, lunar, and stellar spectra. You would then easily understand why Fraunhofer did just this. But his result was only preliminary. I could have told you a story of how, bit by bit, the development led unerringly from Fraunhofer to Kirchhoff. This was, until a short while ago, how most historians of science wrote their histories. But Alexandre Koyré showed in the nineteen thirties that telling stories more like they happened, with the zigzag and false trails, can be more thrilling, and then a much younger historian of science by the name of William McGucken has shown how the result of Fraunhofer was shot down before it was revived by Kirchhoff. In brief, the very able James Forbes, better known as a geologist and an author of travel books, argued very deftly as follows. If the dark lines of the solar spectrum are from the sun's atmosphere, he reasoned intuitively, then they are stronger in the light of the sun's atmosphere seen when the sun is eclipsed.

This is important for three reasons. *First*, the argument is very forceful. *Second*, Forbes and others placed great weight on it. He had a chance to examine his conclusion by observing a spectrum of the sun's corona during a total eclipse; when the moon is between the observers and the sun they can see only its outside, its corona. Luckily for Forbes there was an almost total (annular) eclipse in 1836 in Scotland when he was at Edinburgh University: as the sun was largely obscured its atmosphere could be better seen; the spectrum did not show any darkening of the lines, and so Forbes concluded that the solar absorption spectra are due to earthly atmosphere. At that time experiment could not be sufficiently precise; spectral photography began only in the forties of the century (as soon as photography was invented), and reasonably good spectral photographs were available only in the seventies. Later on Kirchhoff concluded that in total eclipse the sun's corona should not darken the dark lines but, on the contrary, brighten them. And this is the *third* important aspect of Forbes' reasoning: it is quite intuitive, yet erroneous. (The *Dictionary of Scientific Biography* does not mention his contribution, I suppose on the intuitive but false supposition that no error can be an important contribution.) Our intuition is not the highest court of appeal. As Imre Lakatos, the great philosopher of mathematics, said, intuition should develop. The transition from Forbes to Kirchhoff was long and arduous and it is hard to have an adequate idea of the amount of hard work this involved.

I find myself in a very strong conflict every time I read Kirchhoff on the matter. He was an unpleasant and self-asserting pedant yet admirable at the same time. He knew very well that many investigators before him were after the same find, the idea that spectral lines are unmistakable fingerprints for the chemical elements. Yet he claimed priority for this for Bunsen and himself; he announced that nothing will suffice short of a complete proof, the absolute precision of experiment, reasoning, and wording. Making any mistake suffices, in his opinion, to dismiss one's right to the claim of having drawn the much desired conclusion that spectral lines are fingerprints of chemical elements!

I tend to ascribe this find not to Bunsen and Kirchhoff but to Fraunhofer and Talbot, if not also to quite a few intermediaries whose names I shall skip. In favor of this I will say a few things. The importance of the idea of spectral lines as extremely useful—in research and in industry—is unquestionable, and has nothing to do with scientific proof; in this respect it does not matter that Kirchhoff was in error: some spectral lines are molecular, some atomic, and some nuclear, whereas he thought he had proved that all are atomic. Kirchhoff's law, then, is false. Moreover, though admittedly the nineteenth-century quest for scientific certainty is the most ill-advised of all quests for certainty, Kirchhoff at least took it seriously. According to quantum theory, it is likely that, within the limits of accuracy given to any mortal, two elements exist which emit two quite different spectral lines which happen to be of the same frequency. Today no one dreams of identifying an element by one measly line, except possibly in astrophysics, in cases where only one line is available, and then at the risk of a howler. Thus, taking Kirchhoff's own standard as seriously as he wished to be taken, he was not only without proof but also in plain error, and if we take his dismissal of Talbot seriously because of some mistakes, we should dismiss him too, which is absurd: for decades spectroscopy served chemistry better than any other technique and it has done wonders to transform and advance the field. Kirchhoff's contributions were essential to this development. Hence, he was in error when he condemned errors in science so sweepingly. Hence, however, we should not rush to condemn him for his erroneous condemnation of all error in science. As it turns out, his errors were very interesting, and his thinking was both concerning science and concerning radiation. (Positivism, the philosophical claim that scientists have no use for philosophy, at least not in their capacity as scientists, is empirically refuted here.)

These are forceful arguments against Kirchhoff's claim and against his standpoint and mode of argument. Whatever else I say here or elsewhere, I do not wish to withdraw or modify any of them. But I must temper my judgment. I am, almost in principle, an anti-pedant in favor of sloppiness—the principle being that life is too short to spend it on precision. Almost: it is not quite a principle with me; precision at times is a matter of life and death, and minute differences may be amplified to great importance. But how precise should one be? Here Kirchhoff was in error in theory, but right in practice. (The proper application of the right theory assures being right in practice, no matter what they say; but the application of the wrong theory, proper or not, permits being right in practice—whether by luck or due to special circumstances which do not recognize the error involved.) Kirchhoff could not possibly attain absolute precision, or even compete with his followers in matters of precision. He had excellent reasons to attain as much precision as he could, and his announcement of his law, which is the background to quantum theory, possibly would not have taken place but for his passion for precision. Of course, today it is very easy to find imprecision in his work. It is harder to see how comprehensive his passion for precision was, in matters of experiment, measurement, calculation and comprehensive outlook.

My main disagreement with Kirchhoff is philosophical: I follow Whewell and Popper, of not only Popper, and hold a view contrary to his: precision matters not as a means of proof (all scientific theory can be upset regardless of the precision of support), but it is a marvellous instrument of disproof, of overturning present theories. As theories increase the degree of precision, so the experiments designed to test them have to increase in precision. This, say William Whewell and Karl Popper, is why precision of new experiments has to exceed the precision of old experiments; a new observation has to be more precise than the old one as it should be precise enough to present a test of the new and more precise theory.

What good does an overthrow do? It clears the ground and raises the desire for a newer and better theory: the error responsible for the overthrow may serve as a valuable clue: not all errors are important, but some important ideas are errors.

Those who want more tangible results may observe that a confirmed theory, and more so a confirmed residue of a refuted theory, may be taken as a sufficiently solid basis with which to test another theory. Not solid since such things do not exist in the (scientific) world, but solid enough to help shatter another theory; in my view scientific research is like a bootstrap operation where one pulls oneself up by pulling one's bootstraps. Take the discovery of new elements. It is no accident that Fox Talbot could not even think about it—he was too busy with earlier tests. Bunsen took the idea as solid and refuted the standard identification of samples of chemicals from very well known sources considered to be pure sodium or pure potassium; he found them to contain over 30 percent cesium—his new element!

5.2   Absorption Spectra

Kirchhoff's researches involved priority disputes on *three* issues, and on all *three* he claimed priority for a find not on the basis of chronological evidence but on the ground of excellence: his find was valuable because it was the first absolutely exact one.

This introduced a new standard into science, both regarding excellence and regarding priority. I deplore this introduction of a new standard by the way of an example with no general discussion of the general principles involved. The principles smuggled into research a new methodology (science is proof by precision) and a new sociology (science has no room for mere amateurs), and I happen to reject them both. I may very well be in error, but I do not wish my view ignored without a proper debate. It is well known that precedents alone are no substitute for a general debate since a precedent may succeed for a very specific reason. Here is the place to describe and explain Kirchhoff's success.

Kirchhoff's *first* point, we remember, concerned the claim that all discrete emission spectra are purely atomic. The reason for Kirchhoff's success here was sheer luck. In 1862, no more than two years after Kirchhoff argued his case, Alexander Mischerlich claimed to have discovered the first chemical spectra—spectral lines caused by molecules. Also, Lockyer soon showed that metals which give continuous spectra when red hot emit discrete spectra in a spark or an arc as they are then in a much higher temperature. I hope you remember that early in this book I listed sparks as cold light, together with fireflies and cats' eyes. This was a mistake on my part. One of these errors I have already corrected. A short time before Kirchhoff entered the scene, Helmholtz discovered that cats' eyes are no sources of light, you remember. But now is the time to correct my error of describing sparks as cold. How are we to decide the temperature of a spark? Every time you take off a nylon shirt, chances are you create a lot of quite large sparks resulting from terrific discharges; some of them are inches long, as you can easily see for yourself in the dark, especially in a well-heated room in winter (since the room is then exceptionally dry). Yet they do not seem hot as they cause no burns and no sensation of heat. The reason is simple: in order to achieve the sensation of heat and burns, not only high temperature is required but also a large quantity of heat; sparks are very hot but have little quantities of heat. I hope this gives you a reason to appreciate how hard it was to discover that sparks are very hot. My concern here is not the matter of measuring the heat of a spark but the luck that Kirchhoff had in his inability to observe molecular spectra. The considerations of temperature concern not so much emissions from sparks, as sparks emit atomic spectra, but rather the visibility of these spectra. These invite considerations concerning the energy of radiations from chemicals and the energy of molecular emission. The reason is simple: the energy of molecular emission is much smaller than the energy of atomic emission, and, according to quantum theory, the emission of low energy is in the infrared while the emission of higher energy is in the visible light. The emission of still higher energy, including nuclear emission, is in the x-ray region. As this was unknown, the identification of all spectra with atomic spectra was, to a large extent, due to Kirchhoff's attention to visible spectra to the exclusion of infrared spectra. We should not complain about his neglect of the infrared part of the spectrum but be grateful for his study of the visible part of the spectrum; nevertheless, his conclusion that all spectra are atomic is an error due partly to the blindness of the human eye to infrared light—a matter of sheer luck.

Kirchhoff's *second* point concerned his identification of absorption spectra with emission spectra. This identification began with Fraunhofer in 1817, and was accepted by Herschel, Brewster, and many others. Doubts nonetheless surrounded the case. Kirchhoff's self-appointed task was to dispel the doubts. In retrospect the source of the doubts is all too clear: radiation should be a resonance phenomenon, the classical theory said, but the classical theory was in error.

Two great researchers, Foucault and Stokes, tried an experiment to explore resonance. Kirchhoff repeated it. The experiments made no impression, indeed Foucault did not even publish his result. I shall return to them soon.

Before that, the claim that emission and absorption have the same frequencies has to be established. The obvious beginning point for that is to use the same spectroscope for both absorption and emission in order to see whether the lines coincide. Adding increasing quantities of the emission spectrum of the sodium D line to the absorption spectrum of the same sodium D line in superposition should enable us to see solar dark lines get brightened and then obliterated.

Kirchhoff tried this experiment; it did not work that way. The simplest way to perform the experiment, to mix the light from the two sources, is to allow a ray of sunlight to pass through a source of yellow light on its way to the spectroscope. The result was incredible: the dark lines did not disappear; they became even darker. This accords with the (false!) classical theory of resonance: a yellow flame resonates to sunlight of the same wavelength and so disperses it. Admittedly, the flame contributes its own energy, but it also disperses the sun's energy, and, since sunlight is much stronger than the flame, the loss is larger than the gain!

Kirchhoff went one step further though in my opinion this step should have come first as even Fraunhofer could easily think of it. Take a continuous light, let it go through sodium vapors, and put it into a spectroscope. You should now see a dark sodium line!

There is no difference between the two experiments, really, between the passing of sunlight through a yellow sodium flame or its passing through sodium gas. The difference between the two, the sodium vapor and the sodium flame, as Newton already knew, was only a matter of temperature. So here comes the real question: does a change of temperature change the properties of absorption? How? Does anything else contribute to the change? Yes: for one flame to obscure another it has to be colder. Think!

We have arrived at Kirchhoff's *third* point regarding these two sources of radiation. Prevost had said, a body can emit what it can absorb and *vice versa* while cooling or heating up; hence, in one given system, the colder body absorbs more than it emits and *vice versa*. It is very tempting to add that this holds for every wave length, especially the sodium yellow. This, or something like this, was said both by Ångström, and by Balfour Stewart. It happens to be obviously false as we have already seen; a piece of metal emits (everything always emits and absorbs) infrared light in low temperatures and visible light in high temperatures, but still its spectrum is continuous; in very high temperatures its spectrum becomes discrete. Now Ångström and Stewart took care of that. Ångström said that a body absorbs when cold what it can emit when hot. This is false, or at least not clear enough. Stewart said that for a given temperature, what a body emits it can absorb. All this is not sufficient for our purpose of examining the experiment described above as its two flames are not of equal temperatures.

Stewart's argument was simple: take a plate heated to a given temperature and emitting some radiation; cover the plate by another plate of the same matter at the same temperature. Surely the character of the emitted radiation does not change (it was always assumed that the thickness of the plate makes no difference). Hence, the added plate emits on one side exactly the radiation it absorbs on the other, and this holds for each wave length. This argument proves little since possibly a body can absorb radiation which it does not emit. At most the proof shows that what it does emit, it also absorbs; it says nothing about what it can absorb but not emit to begin with. Or does it? A priority dispute raged between Stewart and Kirchhoff. Kirchhoff said Stewart had proved nothing and Stewart called him a pedant: Kirchhoff's proof of his law was much too long and much too inaccurate, but nevertheless he scored. The importance of the proof is not that it told us that indeed absorption is negative emission, but that it raised forcefully the question, why? Kirchhoff drew out of thermodynamics as much as he could; and later on he got help from Wien and Planck in the same direction. The result had to come, as we have seen, yet it did not: and so, finally, the wave theory of light had to be tampered with. But this is hindsight.

I must use more hindsight as I have no intention of giving you Kirchhoff's original cumbersome proof. I shall give you a different argument, more modern and easier. It is based on mistaken assumptions which will be explicitly stated; he made them but without being very clear about them.

Stewart said, a body's emission equals its absorption. But what causes emission and absorption? Kirchhoff observed, and Stewart did not, that emission depends only on the atoms of matter and their temperature—and on nothing else. Hot atoms emit. (Atoms cannot be hot, but let this ride now.) *The ability to emit is the same as emission.* Surely not so with absorption: *absorption depends both on the ability to absorb and on the presence of radiation energy to absorb.*

Here we see a great difficulty which vitiated every discussion up to Kirchhoff (and indeed most modern discussions of Kirchhoff's law; when I was a student I simply could not understand my textbook and blamed myself for its confusion). Emission and absorption are very different processes.

This can be put into two formulas characterizing a given body. First, its ability to emit, its **emission coefficient** as Kirchhoff called the number representing this ability, equals its emission; the emission coefficient characterizes the amount of energy which a unit-surface emits in a unit of time. Its **absorption coefficient**, however, characterizes the portion of incident beam which a unit-surface of it absorbs in a unit of time; the actual absorption equals the actual radiation times the absorption coefficient, and the absorption coefficient can never be larger than unity. It is unity when the body absorbs all the energy that hits it.

Consider the emission and absorption coefficients of a given body. The coefficients are different for different wavelengths: materials emit and absorb different colors with different intensity. Similarly, they do so differently at different temperatures. Thus for a given body its coefficients depend, at the very least, on the temperature of that body and on the colors, or the light wavelengths, that it might emit or absorb. No other factor is involved, said Kirchhoff. (He was in error, but we follow him so as to see where he was going.) What Kirchhoff wanted to find is some mathematical law which correlates the absorption and the emission spectra.

Though Kirchhoff's law is the crown of his success, and the real topic of the present study, I hope you are impressed with what you already know about it. If not, some comparison may help. Remember Foucault and Stokes. What Foucault did is, in retrospect, quite remarkable. He took two sources of light, a large flame and a small arc, and he put the small one between the big one and his eyes. One might expect, under this condition, to see the small light add to the brightness of the large one. In fact the small light looks dark. True, it emits light, but under this condition it absorbs more; brightness is a relative matter as you can see from putting a gray picture once on a white wall, where it looks dark, and once on a black wall, where it looks bright. (Do it; making simple experiments is unbelievably useful. If the classical scientific experience is of any value for us today, then this is it. Even the use of a candle and an electric light permits something like Foucault's experiment, so you can easily perform it too!) Like a gray picture on a white wall, Foucault's small flame obscures the large flame more than adds to it and so it looks dark. Had Foucault worked the idea out in detail, he would have reached the place where Kirchhoff was when he began developing his law. Stokes explained what Foucault saw as resonance, true in a vague way but not very helpful. Stokes must have had his doubts since he did not publish his results; he did not even publish his (true) idea that the yellow sodium light belongs to sodium alone; and, we remember, success in spectral analysis began with this claim!

The main function of absorption spectra, to repeat, was to find what is there in the sun, planets, and perhaps even stars. But only when Kirchhoff, together with Bunsen, made them very useful on earth, could they be used equally well in the sky. Kirchhoff developed a theory of the sun: it had a hard core which was incandescent, perhaps a fluid core, and an atmosphere, the corona, which is a radiating gas. The theory was confirmed in a total eclipse when the spectrum of the corona was taken and, wouldn't you know, the dark lines turned out bright. This was in 1870 and 1871.

Confirmed as it was, the theory is nevertheless false. For the confirmed result was only of Foucault's experiment, no more. This is true of every confirmation: it only confirms a conclusion of a theory, and both true theories and false theories yields true conclusions; the difference between the true and the false theories is in that only the false ones yield false conclusions. Hence, there is no guarantee that a confirmed theory is true. Rather, many philosophers consider confirmed theories probable. To let you know a trade secret, they do not know what they mean by "probable"; confirmation is is better than probability, as it is more comprehensible. (The last opus on probability is by Rudolf Carnap, a leading philosopher of the last generation; it is a thick tome meant to clarify the concept of confirmation by the suggestion that confirmation is probability. It is incoherent and by now practically forgotten.) The following is not contested. *A scientific theory is better the more reported observations it explains and a confirmation adds to their stock, so that anyone who wishes to replace a theory seeks a new one which explains the confirmations of the old one.*

The presentation of confirmation as a challenge to those who do not think the confirmed idea true is only a part of it.

A confirmation may help thinkers develop their ideas: ever so often a confirmed idea is known to be unsatisfactory, yet the confirmation encourages researchers in the pursuit of their ideas. The classical example is Bohr's atomic theory, yet we have here an equally impressive instance: the discovery that absorption spectra come from the sun's corona is far from satisfactory, yet it helped Kirchhoff in his thinking of absorption and emission spectra as the reverse of each other.

5.3 Emission and Absorption Coefficients

Our next step is the search for a formula presenting absorption as the negative of emission. Here Kirchhoff, fortunately for us, made a serious mistake. He wanted proof, and he got some sort of proof of a rather remarkable result, but not of what he really wanted explained, namely that absorption is the negative of emission. This fact, he said, is very strongly confirmed to the extent that it is empirically proven: it is impossible that so many thousands of emission lines, with no exception, coincide so well with so many thousands of absorption lines—they must be negatives of each other. Let us admit this argument though it is rather flat. In any case it is a matter of fact, and a fact is what we want explained theoretically. Let us see where this fact takes us.

For the sake of a proof it is easier to introduce symbols and equations, and I hope you will not mind them: they are as simple and easy as possible, and I shall not assume that you have any knowledge of symbols or of how to manipulate them. What you have to know is that symbols like A, B, C, or a, b, c, or x, y, z, always stand for some quantities: for "A" read "the magnitude A," or "the magnitude of A," or "the quantity A," or "the quantity of A"—at your convenience. Further, the equation

$$A = A(B, C).$$

says, oddly, I am afraid, that the quantity A depends on the quantity B and the quantity C alone. For example, when A is the cost of a basket of apples, B is the price of an apple and C is the quantity of apples in the basket, you can read this to say "the quantity of money paid for the basket of apples depends on the price of an apple and the number of apples in the basket alone". You can further write,

$$A(B, C) = B \times C,$$

and read this to say "it equals the product of these two numbers". You can write the two equations together, merely in order to save space, and it will look this way:

$$A = A(B, C) = B \times C,$$

and read this the same way as before. It may sound odd that we first say that A depends on B and on C alone, and then say that the dependence is a product of these two numbers; this, however, is quite natural when the right formula is negotiated. For example, we may say we do not know what price to charge you for the purchase of the commodity in question, but the price should be clearly dependent on the price of a unit of that commodity the number of units of that commodity purchased, and on nothing else. If the price depends on these two factors and other known or unknown ones (say, the purchase is of a perishable commodity), we indicate our ignorance by writing

$$A = A(B, C, ...).$$

to read "the magnitude A depends on the magnitudes B and C and on other unspecified magnitudes". If a dependence on some magnitude (say, C) turns out to be of no import, we note this in the following manner:

$$A(B, C) = A(B)$$

and read this to say, the magnitude that depends on two other magnitudes turns out to depend on one of them alone, or something to that effect. If we have only

$$A = A(B)$$

and *B* makes no difference either, then we write

$$A = A(B) = constant.$$

It is funny that dependence can be called constancy (which is really not dependence at all) but mathematicians talk this way. Wishing to ask, does *A* depend on *B* and if so how? they ask, how does *A* depend on *B*? and they take independence to be the answer, in no way. It is a mere matter of convenience of mathematical expression. This will do for now.
Call the emission of a body *E* and its absorption *A*.
Call its emission coefficient *e* and its absorption coefficient *a*.
(Remember: the absorption is of a fraction of the incidental beam, and so the absorption coefficient is a fraction.) It was Kirchhoff who postulated the idea that emission is the emission coefficient represents emission. Erroneous as this idea is, it is one of the most challenging and significant in the history of physics; it had to be conceived before one could suggest the contrary. The importance of this idea can be put like this: nuclear piles and lasers came as refutations of Kirchhoff's idea; hence, no Kirchhoff, no nuclear piles and no lasers.

Let me make a small comment here. The view that actual emission equals possible emission, that emission equals the emission coefficient, $E = e = e(\lambda, T, ...)$, means the same as the view that emission is totally independent of the surroundings of the emitting body. This is a striking suggestion. We know of almost no other such claim. Nuclear physics shook our philosophy of nature very badly because it suggested that whether and when there will be an explosion of a given nucleus of radioactive matter, such as an atom of radium or of uranium, is totally independent of the environment and predictable only statistically. As you must know by now, even if you have never read anything about nuclear physics, this claim is tragically false (at least tragic for the inhabitants of Hiroshima and Nagasaki, but surely for the whole world): there are devices designed to manipulate the statistics, to increase the rate of disintegration of radioactive nuclei; these devices, in other words, raise dramatically the probability that the radioactive atoms in a nuclear mechanism will disintegrate, rather slowly if the mechanism is a nuclear plant and frightfully fast if it is a nuclear weapon. How is this accomplished? It is done quite indirectly. Let me explain this by giving you first a false theory of how it is done and then a theory presumed to be true.

First, false option: a radioactive nucleus is unstable and can disintegrate at any time. When hit with some missile, its chance to disintegrate will significantly increases. A missile that can hit a nucleus has to be its size, say, a neutron. The neutron has a chance to hit the nucleus and thus hasten its disintegration.

Second option: a nuclear device alters the nucleus which it tampers with. The neutron does not make the nucleus it hits disintegrate; rather, it enters the nucleus. The nucleus then becomes a different kind of nucleus with a different disintegration probability.

This idea should not surprise you too much: its roots are in an idea which was suggested by Prevost, no less. It was Prevost who said, you remember, that when a body is illuminated, when it is bombarded with radiation or light energy, it heats up, and it radiates back because it is hot, not because it is bombarded with radiation (except, of course, that the analogy between radiation and radioactivity is very partial).

Moreover, I do not know if it is true of nuclear forces in general that they are not directly controllable, that they are not directly inducible (if I may use a new technical term), that they cannot be bombarded so that they disintegrate under the sheer impact rather than as a result of a change. But this is unimportant for us as the present story is limited to the emissions of light by atoms. Prevost and Kirchhoff and many, many other thinkers maintain that radiation cannot be induced. They are in error. In the early twentieth century Wood and others succeeded in performing experiments with induced emission. Without knowing about this, Einstein developed a theory of induced emission in 1916. (This theory was soon over-shadowed by Bohr's and his followers' work which was soon crowned with the development of quantum theory in 1926. Induced emission, and Einstein's study of it, lay dormant for a few decades until instruments that operate on induced emission were created (lasers), a very exciting development.) But no induced emission for Kirchhoff.

I hope you are not impatient with me. I had to take this slowly as it took me years to work it out. Though I studied physics after Hiroshima, I was still taught Kirchhoff's law as if it was the gospel truth. I debated with others at great length the marvellous idea that the atom disintegrates according to its own statistical laws regardless of the environment, thus proving that they escape the laws of causality once and for all, yet we did not notice that this conflicts with Kirchhoff's law. According to Kirchhoff's law emission equals the emission coefficient. If you wish to change the emission pattern of a body, you must tamper with it; but you cannot tamper with its emission mechanism. To put it in another terminology, emission is a function of internal variables only. We write this thus:

$$E = e = e(\lambda, T, \ldots)$$

where E stands for a body's emission, e for its emission coefficient, and its emission coefficient depend on the two factors, color or wavelength, denoted by the Greek letter λ (*lambda*), and temperature denoted by T, and other, unspecified internal factors, denoted by the dots, but by no factors characteristic of the environment.

$$A = a \times I$$

is the equation for absorption of any given body where A is its absorption, I is the strength of the light beam that falls on it (both per area unit), and a is the fraction of the energy of the incident beam that is absorbed; now whereas the incident beam depends on the surrounding, the absorption coefficient does not:

$$a = a(\lambda, T, \ldots)$$

where again the dots stand for internal factors only.

We need not go into this matter, but since Kirchhoff did, it is traditional to discuss, however cursorily, the relation between the light in the environment and the incident beam; we may shine a strong light on a body and yet the strength of the incident beam may be minimal since light may hit a body at a very acute angle (as sunlight hits our environment at sunrise and at sunset), and the strength of the incident beam may be maximal since light may hit a body perpendicularly (as sunlight hits the environment at noon). Thus, for the incident beam, its strength I is

$$I = I\,(\,\lambda\,,\,\textit{intensity of the given light beam, direction of its incidence}\,)\,,$$

meaning that we take as given the distribution of light in the environment, its intensity, and its angle of incidence, and we consider the strength of the incident beam as dependent on these and on these alone. The details, to repeat, do not matter; what matters is merely the obvious dependence of absorption on the incident beam I:

$$A = a \times I.$$

Clearly, the emission and the absorption coefficients are totally different kettles of fish. Both emission and absorption are of energy per unit of time, but as the incidental beam is energy per unit of area and unit of time, clearly the absorption coefficient is not a coefficient of the same quantity as absorption; there is the matter of the direction of incidence. For example, when the incidental light is a ray of sunlight, the angle between it and a line perpendicular to the body the ray hits (this perpendicular is called the normal) makes a great difference, just as sunlight is strongest at noon and weakest at dawn or dusk. We want the full value of the energy to be considered when the light ray is parallel to the normal and to vanish when it gets perpendicular to the normal. (The dependence on the direction, then, should make the incident beam have full effect when it is perpendicular to the surface of the body in question and zero when parallel to it; this is the projection of the intensity of the incident beam on the normal, or the product of the initial intensity of the incident beam with the cosine of the angle of incidence.)

Emission looks much simpler if it is the same as the emission coefficient which depends on color and temperature alone. This claim turns out to be false; remember induced emission discovered by Wood empirically and by Einstein theoretically. Let this illustrate what I consider a great lesson: great scientific discoveries, as Galileo has observed, lurk in clear thinking and the correction of some current errors.

5.4 Kirchhoff's Law

The empirical fact Kirchhoff thought he had proven is this: *whatever wavelength a body cannot emit, it cannot absorb, and vice versa.* It is a law of resonance; for all we know it is true, and it is the kernel shared on the matter by all investigators. It is often called the law of reversibility or of symmetry. It is still a sacred law of quantum theory and many a revolution was effected within quantum theory with the aid of this law. Written in mathematical symbols this may look thus:

$$e = 0 \quad \text{if and only if} \quad a = 0.$$

This same idea can be more elegantly written as,

$$e = k \times a \quad \text{and} \quad k \neq 0,$$

that is to say, k is some non-zero quantity. (Being an absorption coefficient, a is between zero and unity: $a = 1$ means the absorption of all incident light, and $a = 0$ means no absorption at all.) When $a = 0$ for all wavelengths and temperatures, we write $a \equiv 0$, and read "a is identically zero", where its being identically zero means that it is zero for all wavelengths and for all temperatures; bodies with this characteristic are perfect mirrors for all wavelengths and for all temperatures. There is no perfect mirror. (To be precise, a body for which a is zero is a perfect mirror or perfectly transparent or a perfect semi-transparent mirror. Yet we systematically ignore transparency here since we are concerned with the interaction of matter and light and transparency is no interaction.) Similarly, $a \equiv 1$ for a given wavelength and temperature means that the body with this characteristic is a perfect absorbent of that wavelength at that temperature. For a perfect absorbent of all wavelengths in all temperatures, $a = a(\lambda, T, ...) \equiv 1$; it is perfectly black. There is no perfectly black body.

$$e = k \times a; \quad k \neq 0, \text{ and } k \text{ is a positive number}.$$

$$e = k \times a; \quad k > 0,$$

$$e / a = k; \quad k > 0.$$

More fully,

$$e(\lambda, T, ...) / a(\lambda, T, ...) = k(\lambda, T, ...) > 0;$$

the dots stand for unknown internal factors, not external ones.

It is worth noticing again that e depends on wavelengths and internal factors since it is an emission coefficient, just as a is, and likewise k depends on the same factors as it turns out to be the absorption coefficient of any black thing. The bold hypothesis is that E depends only on e, which is the assumption that emission is not induced, that emission depends only on the state of the emitting atom—though an atom can emit only if it stores some energy that it can discharge, of course. Such an atom is called excited; and an atom is excited because it has absorbed energy from the environment. Thus, though the emission coefficient represents the environment in some sense, the idea that emission equals the emission coefficient severely restricts the manner in which the environment can cause emission; in particular, it cannot induce emission. The discussion is interesting even though this hypothesis is false since whenever it holds, k turns out to be the correlation between the emission and absorption coefficients of any substance. It is assumed throughout the discussion of radiation theory that the emission mechanism of an atom and its absorption mechanism are identical, and this is the central assumption that quantum mechanics takes over and sustains to this very day. It stands to reason, then, that if the magnitude k which depends on color and temperature alone is the only way to translate the description of absorption to that of emission, we may prefer to study absorption since the absorption coefficient is a magnitude mathematically easy to handle as it is between zero and unity, whereas the emission coefficient is the actual emission which may have this intensity or that. In any case, k is obviously quite intriguing as it is allegedly a universal factor relating heat and light and depending on color and temperature alone.

In the pursuit of information about k we can consider its conduct under some external factors congenial to us since this will not alter k as k is independent of all of them. That is to say, whatever holds for k under any given external circumstances holds for k in general. This is terrific because to think of a system in equilibrium is the easiest—particularly in thermodynamics where all we know is really about heat in equilibrium; we know only a few facts about heat-flow.

What we want, first and foremost, is to compare k for different bodies. One reason is really very simple and technical. For a perfectly black body, $a \equiv 1$, the law $e/a = k$ becomes $e = k$. This holds for all "black" bodies! This sounds strange, since a black body radiating is not blacklooking. This puzzles novices and so they are reminded (by the more sensitive teachers and authors of introductory textbooks) of the fact that charcoal is black when cold and it glows when hot. This clarification is not sufficient as coal is supposed to be black in all temperatures! It is: it absorbs all the light that hits it, but as it glows this fact cannot be observed; careful experiments (of the kind described at length in a previous section) by Foucault and Kirchhoff, did show that while glowing live charcoal devours all the light that falls on it. Or almost all, as no matter is perfectly black, not even charcoal.

It is startling that at the same temperature all black bodies emit the same light. It is not empirically startling, since at most one or two kinds of black body are available, charcoal, or rather soot, and platinum black, which is thinly pulverized platinum, and even they are not quite black. It is theoretically startling: we began by saying that emission patterns are characteristic of the emitting atoms, and now we say, any atom of any black body emits the same light at the same temperature. (What is the temperature of an atom? can there be such a thing? It is hard to say, but do not let this worry you just now: one thing at a time.)

On second thought, perhaps it is theoretically not so startling as it amounts to saying that there can be no black atoms. This renders the present discourse entirely fictitious. It is not: it is the significant claim that whether black bodies exist or not, their radiation is the ratio between the emission and absorption coefficients of all atoms; hence, even were there no black bodies, that ratio is claimed to be universal.

Back to the magnitude k of different bodies then. How do these magnitudes k of different atoms compare? Take two pieces of different matter, and use them as walls for a cavity in thermal equilibrium. Each wall emits and absorbs its share of radiation, but the system is in equilibrium. In equilibrium each body absorbs as much as it emits:

in equilibrium $\quad E = A$.

(Otherwise there would be a temperature change and thus no equilibrium.) In equilibrium the law of entropy, the second law of thermodynamics, demands that radiation energy should be the same in all directions. This is a terribly important assumption which Kirchhoff did not make. Indeed, Kirchhoff himself knew that radiation need not be in total disorder. He therefore tried to prove his law for every single beam of radiation. His calculation became complicated and thus suspect (all sorts of assumptions may enter complex calculations quite unnoticed). Moreover, all of this is quite besides the point. Since k depends on internal variables alone, we can choose any external condition we like: we can, as a matter of course, assume complete thermodynamic equilibrium, namely, that there exists an enclosure, a cavity, which is:

    a. insulated from the rest of the world;
    b. made of walls of different materials;
    c. in constant temperature;
    d. full of radiation distributed equally in all directions.

Under these conditions, each part of the cavity emits as much as it absorbs. Put into formulas, this reads,

$$E_1 = A_1 \quad \text{and} \quad E_2 = A_2 \quad \text{for bodies 1 and 2}.$$

Under these conditions radiation from every side, from every wall, is equal to that from another wall even if they are made of different materials. Put into formulas, this reads,

$$A_1 = I \times a_1 \quad \text{and} \quad A_2 = I \times a_2 \quad \text{for bodies 1 and 2};$$

$$a_1 = A_1 / I \quad \text{and} \quad a_2 = A_2 / I \quad \text{for bodies 1 and 2}.$$

Since in a cavity in equilibrium, I is the same everywhere,

$$E_1 = e_1 \quad \text{and} \quad E_2 = e_2; \quad k = e_1/a_1 = e_2/a_2 = I.$$

This is extremely comforting since the difference between emission and absorption depends entirely on the intensity of the incidental beam, and k, the factor of proportionality between e and a, is just that intensity in the case of thermal equilibrium.

What characterizes I here is just the temperature of the two bodies and the wavelengths in question; the atomic characteristic of either of the two bodies does not come in at all. If you doubt this claim, you can just as well imagine that one wall of the cavity is made of a black body, where, we remember, $k = e$ and has nothing to do with the specific emission mechanism of the black body, yet $k = I$ there too. Remember that e_1 as e_2 does depend on the atoms emitting light, but not so e_1/a_1 or e_2/a_2 or k. Thus, in thermodynamic equilibrium,

$$k(\lambda, T, ...) = k(\lambda, T).$$

That is to say, under the equilibrium conditions specified, k does not depend on any internal factor, on any atomic mechanism; but since k does not depend on any external factor either, the specified conditions can be ignored.

        k is independent of any conditions;
        k is the same for all bodies in equilibrium;
Therefore,   k is universal.

This is Kirchhoff's law: k is universal.

Since e/a is the same for all atoms, it follows that, at a given temperature, if the emission of a body in some given region of the spectrum is largely confined to the narrow neighborhood around some one or few fixed wavelengths (emission spectral lines), then on the same spot, at the same temperature, the same body also has the same absorption characteristics (absorption spectral lines), and the ratio between the intensity of the emission and the absorption of the body at the given wavelength equals the intensity of the radiation which a black body emits at that wavelength. Where the spectrum of a body is continuous the same holds. Also, for a given wavelength and a given body, if raising its temperature to a certain level brings out emission, then the same temperature will be responsible for absorption there. This is also contrary to Ångström's hypothesis which says, we remember, that absorption occurs at lower temperatures than emission. Ångström's hypothesis is more intuitive since at ordinary temperatures most bodies around us absorb (otherwise they would not seem colored), yet they do not seem to emit the wavelengths they absorb (otherwise, it seems, they would be white). It is counter-intuitive, as we cannot see an emitting body radiate; but it was checked experimentally. A typical example is the charcoal we have met already which seemingly emits nothing and absorbs all when at room temperature but which, in truth, always emits and absorbs. By the Stefan and Boltzmann law, discovered about 20 years after Kirchhoff's law, everything emits all the time except at absolute zero. Absolute zero is never reached. (This is known as the third law of thermodynamics.)

Under the defeatist influence of instrumentalism many writers identify a theory or a law with a formula. In the present case this is rather embarrassing, as Kirchhoff's formula is but of a historical significance, whereas his law is more basic and it is the statement that the formula holds because the mechanism involved in absorption is the same as the mechanism involved in emission. Otherwise the one would not be the negative of the other; the mechanism will not leave the picture when the ratio between them is considered. Though Kirchhoff's law is false, for all we know, this part of it holds: the same mechanisms work for both absorption and emission. All emission-absorption processes as far as we know are reversible; the two processes mirror each other.

What is false in Kirchhoff's claim is not only the stipulation that there is no induced emission. It is also that the emission and absorption mechanism for one wavelength has no effect on the emission and absorption mechanism for another wavelength, that all that matters is the distribution of wavelengths (i.e., one atom emits mainly yellow light, one emits mainly red, and so on). We know that this is not true: one atom, any one atom, can have quite a few electrons that can emit, but usually only a very small number of electrons is involved in the process even if the atom has many electrons. While one electron is busy emitting one wavelength, it is absolutely too busy to do anything else. The same goes for absorption.

Wiedemann, a physicist who was once well-known and is now almost entirely forgotten, younger than Kirchhoff but older than Planck, listed a number of phenomena where the radiation of a body depends not only on its temperature and inner mechanism; he called them all by the collective name of luminescence. He also found deviations from Kirchhoff's law which could be corrected if the numerical value of the body's temperature is slightly altered; Wiedemann called this numerical value "color temperature". It still plays a significant role in astrophysics. The name "color temperature" comes from the fact that Kirchhoff's k correlates wavelengths or colors with temperatures; for a black body, of course, and so for all bodies.

Similarly, the law $A = I \times a$ says, absorption depends only on the mechanism of absorption and the intensity of the incidental beam ; there is no such thing as saturation. This is nearly right because usually emission follows absorption almost at once. But this is not quite so in the case of phosphorescence. The time lag is central to the case of a laser where energy is pumped into a piece of matter for a while and then, due to induced emission, they all fire together in phase.

Paschen and Wiedemann refuted Kirchhoff's law experimentally; they measured the intensity of emissions and absorptions of spectral lines and compared their ratio with their equivalents in a black body's radiation (i.e., the part of a continuous spectrum of the same wavelength).

5.5 Kirchhoff's Followers

A few things personally surprised me when I started studying the original literature on Kirchhoff's law. One of them is its fuzziness. I was naive, of course; one must, in all fairness, expect some lack of clarity in every original contribution. One also expects followers to clear up the mess. Whatever an original contribution is, it usually gets cleared up, at times sooner, at times later, depending on the difficulties involved. To my surprise, Kirchhoff's law is still not cleared up. Another thing that surprised me no less was the apology that Planck and others made for their great interest in Kirchhoff's law. I shall return to it in the end of this chapter. They said that it was the universality of the law that turned them on. This is odd: first, scientists seldom defend their interests, especially when they succeed; second, as all laws are presumed universal anyway, this is no distinction of Kirchhoff's law. Many years ago I raised this difficulty in an essay on the Kirchhoff-Planck law (Appendix A below). My essay was noticed but the difficulty was not. Let me restate it and then elaborate.

The status of Kirchhoff's law is still unclear. When facts turn up which run contrary to an accepted theory, the theory does not usually drop out of science textbooks, nor does it appear there as true. Historians of science often present a refuted theory (say, Newton's or Lavoisier's) as true; this is not the case with the better science textbooks. These usually modify the law, both by slight improvements and by proper qualifications. The modifications are hardly avoidable: things usually first appear messy and their further study, says Karl Popper, invites simplification; simplification, he adds, is usually distortion. (The two words "simplification" and "distortion" have contrary flavors, but ideas are more interesting than the flavor of words.) As a twenty page proof of Kirchhoff's law as given by Kirchhoff, for example, is put here in one or two pages, certainly this involves a distortion, particularly since the original mathematical tools are avoided here. Qualifications differ from simplifications: they are historical distortions aimed at retaining the (presumably) true part of a refuted law. The standard example for this is the classic gas law. It concerns the elasticity of all gases. In its original wording it is known to be false, and so textbooks describe it with the proviso: this law is obeyed by ideal gases and a gas is ideal if it obeys this law.

Notice that this is a circularity that makes the law logically true. It thus invites the question, which gas is ideal (or near-ideal)? No gas is ideal but hydrogen and common air, for example, when not too cold and not too hot, are nearly so.

Not all qualifications need be circular. The standard example for non-circularity is Newtonian mechanics. In the light of Einstein's relativity we restrict it to small velocities and weak gravitation and consider it merely approximately true. But it should also be qualified to exclude quantum mechanics. This is easier said than done. It is repeatedly said that Newton's mechanics holds where Planck's quantum of action (which you do not need to know anything about right now except that it is exceedingly small) can be viewed as zero. This is very unclear. The yellow emission of salt turns out to be the sodium yellow D doublet which is a quantum mechanical affair. Newton's mechanics should be qualified so as to exclude the mechanics of atomic radiation, quantum mechanics, and it is an open question how this should be effected.

A similar, more murky situation occurs with Kirchhoff's law. A very serious person who is a scientist, a philosopher and a friend of mine, found my essay on Kirchhoff quite redundant but changed his verdict (he was the editor's referee) when I added to it a note which contains a list of different qualifications to Kirchhoff's law taken from current leading textbooks and handbooks of physics.

Some say that Kirchhoff's law holds only for thermal radiation. What is thermal radiation? This is hard to say. Originally thermal radiation involved only heat and light; if electricity is involved (photoelectricity) or chemical action (photochemistry), then even Prevost's law does not apply, let alone Kirchhoff's. But now we are convinced that heat does not cause matter to radiate; rather, heat excites electrons bound to matter and then they radiate. This is why we prefer vacuum tubes where free electrons run amuck and bump into the rare atoms of gas that are still in it (no vacuum is perfect) and then the electrons that bump into the atoms excite the electrons which are bound to these atoms, so much so that the whole vacuum tube (which is very cold, of course) glows strongly with very little heating. It is the heating that is energy loss (particularly in incandescence, such as in most light-bulbs). Now, some of those who say that Kirchhoff's law holds only for thermal radiation are obviously sensitive to this. They define thermal radiation not in the traditional way; they say, it is the radiation for which Kirchhoff's law holds. This makes the law logically true, but we should not sneer at it: the right response to such a law is to ask for a list of cases in which it applies. Is there a list of cases for which Kirchhoff's Law applies? I could not find one.

Others qualify Kirchhoff's law to black bodies in thermal equilibrium which is silly, as under such restrictions it follows easily from Prevost's law of exchange that preceded it by almost half a century. Moreover, surely black body radiation is thermal, and so the equilibrium condition is an unnecessary restriction. Some suggest that, since the proof is given for thermal equilibrium, it should hold for all bodies in thermal equilibrium. But, again, the force of the proof is just the claim that, since absorption is atomic, what holds for it in equilibrium holds for it generally. Some respected texts present the law as true without qualifications. In texts on phosphorescence and luminescence, phenomena which obviously disobey the law, naturally the law comes qualified. And in texts on astrophysics—and I hope you remember that since Kirchhoff's law helps convert absorption spectra to emission spectra, it is very useful in astrophysics, of course—the law is presented as a first approximation to better variants all of which are in the wake of Wiedemann's pioneering work at the end of the last century. Around Wiedemann's time the situation was not too clear, and it has not been clarified since. It was clear already then that Kirchhoff's law is not true as it stands, that it holds only under certain qualifications and as a mere approximation. Yet Kirchhoff's great followers, including Planck, presented it as true. This may explain their odd justification of their interest in it; they held a false view of scientific method and demanded that a false theory be discarded and replaced by a better variant; before taking it seriously they wanted to modify or qualify or correct it, though clearly in order to correct a theory one has to take it seriously. Just as great artists are compelled to break established rules of their art—even if it makes them feel guilty—so did these great researchers. This comparison is nice but also too facile to be satisfactory: it lacks the specification as to when and how the rule is broken with impunity—or reformed or replaced!

It is odd that Kirchhoff's law was, and still is, endorsed with no qualifications in spite of such things as cold lights, namely, fireflies, etc. There is an explanation for that (there always is, of course). In some cases, such as the case of the firefly, the law does not apply as they involve energy exchanges other than between heat and light. But not always. In the cases in which the law fails it may perhaps be restored by replacing the temperature T by a similar scale, and then Wiedemann called this scale the color temperature of the case in question.

Is this true? I do not know. Great experts in the field (e.g., Wien in his Nobel Lecture of 1920) say that it is. I doubt it on the ground that the very idea of emission and absorption coefficients is erroneous anyway. But I do not know. In elementary school my science teacher told me, never believe anything except what you see for yourself. This cannot be done. In university courses my professors tried to illustrate as many experiments as they could, describe well those they technically could not, explain well each theory and argument they wanted their students to endorse, etc. Nevertheless, they repeated a lot of opaque material endorsed on trust. At least I felt pressed to endorse, and reluctantly I did. I later found out two things. First, there was a lot of apologetic material there; what they preferred to push on us rather than explain to us was not really capable of a good explanation and defense. Second, I learned from Karl Popper that science is not a matter of consent but of critical disputes; in science only repeatable and repeated observation reports are compulsory to some extent.

Nevertheless, let us suppose that every radiation has a color temperature. This would be a great boon. For, you remember, the law of entropy is the law that disorder is always on the increase, and the concept of entropy or disorder is much broader than the concept of temperature. Since there is a fundamental law of gases which correlates entropy, energy (or quantity of heat), and temperature, it would be terrific if we could have a color temperature for every radiation. The mere thought should offer us a glimpse of a possible solution to our problem. We have a cavity with a fixed temperature, its walls radiating and in equilibrium with the radiation in the cavity. It is tempting to say that in equilibrium the radiation in the cavity has the same temperature as the walls. This ascribes of temperature to *all* radiation. Is this inviting trouble?

The fascination of Kirchhoff's law is understandable. Writers on spectroscopy in the last quarter of the last century expressed great excitement about the law. They defended it by showing how much it had achieved. This is too smug to count: it is little more than crowd pleasing. What is exciting in Kirchhoff's law is a mystery and a promise. It does not make explicit but hints at a view which is now still endorsed:

atomic emission and *absorption mechanisms are identical*.

Questions crowded the scene: how do light and matter interact? How do light and heat? And electricity? What mechanisms do atoms have for interaction with light?

It was all breathtaking and they worked feverishly on diverse fronts. One front, in particular, was the question, what is the function k? It could be measured since it was the distribution of radiation in a black cavity; and we can have coal cavities heated to different temperatures, canals dug into them, closed with filters to transmit monochromatic light to be converted by a thermocouple to a current measured by reading the position of a needle on a sensitive galvanometer. The experiment is devoid of theoretical snarls, a relatively straightforward if arduous job. A feverish competition went on among the few who felt equal to task of harmonizing theory and experiment.

5.6 Spectral Lines Between Kirchhoff and Bohr

Kirchhoff's law came to throw light on absorption spectra and so serve astronomy. It became useful for chemical analysis too. But it posed a very challenging theoretical problem, in particular, the problem concerning the exact ratio between the absorption and emission coefficients. Now this question, according to Kirchhoff's theory, can be translated into the question, what is the profile of a black body radiation? how does its intensity, in other words, depend on its temperature?

This question diverts interest from spectral lines. The growth of radiation theory from 1860 onwards, as outlined in almost all histories and introductions to quantum theory, is described as focusing on this question: Boltzmann (1879), Wien (1897), Planck (1899), Einstein (1905), Bohr (1913). None of them except Bohr (1913) cared much about spectra, and Bohr told some interested historians that he developed his ideas thinking only about the work of Planck and Einstein on black-body radiation and of Rutherford on atomic structure, and only when his ideas were at a fairly developed stage, did he search for applications for his theory, and, naturally, it dawned on him that his theory might explain atomic spectra.

This is an almost incredible coincidence, except that it was no coincidence at all. For the central problem always was, how can an atom radiate? This, more than any other question, led to the suspicion that the atom is not that unsplittable. I really wonder if the stories about researchers before Rutherford being so convinced of the inability to split atom are not myths.

Come to think of it, we are obviously too unfair to our predecessors: we mock those who did not believe atoms exist, and we mock those who said atoms are truly atomic (in Greek "atomos" is unchoppable or unsplittable or indivisible). We introduce much muddle by calling a splittable particle unsplittable and then blame everyone who is reluctant to join us in our narrow escape from it. Also, we should notice, Bohr's hydrogen atom made up of one proton and one electron is, clearly, a split atom. It was not the first split atom, but it was the first highly detailed theoretical split atom.

The nineteenth century theorists had hoped to describe radiating atoms as resonators giving series of spectral lines like vibrating strings. Their dream came true in the mid-twenties when Louis de Broglie and Erwin Schrödinger developed wave mechanics. This could not be done in the nineteenth century when mechanics was Newtonian and its matter indestructible and when charges were without electrons.

Still, the search was at least linked to atomic spectra and clearly had great success even if it led to no satisfactory theoretical results; it encouraged experimenters to try and find some order in the chaos of the mounting thousands of spectral lines. I should refer you here again to William McGucken's book, *Nineteenth Century Spectroscopy, 1802—1897* (1969), for more details. For my part I shall skip most of the story and only mention the facts leading up to Bohr.

The search for order was a search for series of spectral lines. These series are harmonic if, indeed, the phenomena of spectroscopy are those of classical resonance. We do not know if spectra represent resonance. Erwin Schrödinger's classical paper, "Are There Quantum Jumps?", which is his credo and *cri de coeur*, presents resonance as at the root of quantum theory. But he represented the minority view. And he stressed that the laws of quantum resonance are far from being classical.

This matters little for our narrative. A search for a series may be called literally a search for harmony, be it of resonance or of anything else. Alternatively, calling a quest a search for harmony is metaphorical, like the old Pythagoreans who spoke of the harmony of the spheres, or as Kepler who looked for a series relating distances of planets from the sun and who also spoke of the harmony of the spheres. (Perhaps the case is the other way around: as harmony comes from the Greek verb "harmozein", to fit together, perhaps harmony in music is but one case of fitting together; and Kepler's search was for another, equally legitimate fitting.)

It is hard to assess such searches. They involve a lot of gut-feeling. Strictly, any mess can be organized in some series, provided the series are allowed to be messy enough. This fact often confuses people not well-versed in science, especially when they think that science ought to look for the wondrous rather than for the simple. But only when we have a body of theory and rules which are based on the theory and which generate a series of numbers, only then can we say that the series are more messy or less messy, more arbitrary or less arbitrary. And so some arbitrariness has to be allowed the researcher since it may or may not be vindicated by later theorizing. (Scientists, said Einstein, are opportunists.)

This is why I wish to skip this admittedly fascinating story. I know that all detective novelists remind their readers how much legwork a true case involves, how many times the hero tries in vain to fit together the different pieces of the jigsaw-puzzle, especially since the chief pieces are missing and wrong ones are thrown in for good measure. But novelists only say so; they do not put you to the task of reading dozens of pages describing the tedious legwork and fitting together. They (the detective novelists) give their readers the feel of detectives doing their investigative routine work. They (the detectives) may be young assistants with no interest or concern, but forced to do their jobs tolerably well while feeling dead inside. Or they may be burning with concern (the victims were colleagues or best friends) or with ambition (I must crack this case and show them, the detective mutters) or with genuine interest and almost scientific delight (a veritable Sherlock Holmes).

I am now omitting a chapter on the tedious work that went on for decades in efforts to put some order into the colossal mess that is the list of all the spectral lines known. What I like about the chapter I am omitting is that it is full of all of these characters and more. It is also a chapter supposed to be laced with numerology. Numerology is on the borderline between science and superstition. Many amateurs who know almost nothing about astronomy know all the facts about the nonsensical Bode's law which is a numerological relation between the planets' distances. But the terrific thing is that some of the series did work: Balmer's and Rydberg's. Balmer was a Swiss school teacher; his rule was developed by Rydberg who later became a professor of physics in Stockholm. His result turned out to be a part of Bohr's system. I do not know what considerations these two thinkers had, if any, and if they really were numerologists; they are not discussed in the literature, and the suggestion is that they were numerologists of sorts. Someone might wish to check these things.

Bohr worked on both the offshoots of Kirchhoff's law—Planck's and Einstein's—and on the offshoots of models of radiating atoms. Quite a few top researchers tried their hands in the project: Helmholtz, Kelvin, J.J. Thomson who had discovered the electron, and Ernest Rutherford, Bohr's teacher. And so, in Bohr, all three trends united and put an end to our story; this is no accident since all three trends have a common origin in Fraunhofer's absorption and emission spectra.

It is no accident either that though Planck is viewed as the father of quantum theory, the birthday of which is fixed as December 14th, 1900, the old quantum theory proper is universally considered to be not Planck's theory of that date, nor Einstein's theory of 1905, but Bohr's theory of 1913. This was, indeed, the end of radiation theory and of classical atomic physics, the beginning of the old quantum theory.

Allow me one paragraph for a minor point with a moral; it is unrelated to the story at hand, but you may find it useful. Any monochromatic light is characterized by its wavelength λ and frequency ν and it does not matter with which the light in question is characterized, since $\lambda \times \nu = c$ and c is the speed of light, a universal constant (about three hundred thousand kilometers per second). But actually, ν is a bothersome entity. λ is the length of a wave, the distance in units of length between two adjacent wave-crests. Corresponding to it should be the distance in units of time between two adjacent wave-crests, not frequency. Looking at a wave front at one point, measurement can be taken of the duration of the passage of two adjacent crests through it. If it is a standing wave, such as on a violin string or a drum's skin or water in a tub, measurement can be taken of the duration of the passage of the string or skin or water down and up again at one point. This unit of time is called the period T, and it makes much better intuitive sense to view the velocity of light, c, as the ratio between a distance and the time it takes light to traverse it since velocity is by definition the ratio of distace traversed and the time it takes to traverse that distance. There is no objection to the use of $\nu = 1/T$ and so, $\lambda \times \nu = c$ is the same as $\lambda / T = c$. But novices are puzzled as to why the physics teacher uses ν instead of T. The reason is simple: for many purposes (not all) ν is more convenient than T. But novices do not know that and are puzzled and bullied to accept ν rather than T. Also, just as we have ν as the number of waves per unit of time, so we can have the number of waves per unit of length, the so-called wave number, designated as k where $k = 1/\lambda$. (Do not confuse the k in the radiation formula with the k here: the one is the function characterizing the spectrum of a black body, the other is a wave number characterizing a given monochromatic wave.) So we can characterize any spectral line by k or by ν or by T. When in the dark, you try every available possibility. It so happened that some time in the eighteen seventies it turned out that it was easier to show some regularities in spectral analysis by using k. Until the advent of wave mechanics in 1926, it was not clear why. Now it looks so obvious that teachers start imposing k on students the way they used to impose ν on them, so as to make them feel used to it as a matter of course and just out of habit.

The view of science as trial and error (as announced by Karl Popper) is now increasingly popular. In a way there is no denying it: mistakes are always possible, and corrections should always be welcome. Yet as a characterization it is too concise, as it fits both the spectroscopist's search for a formula that looks too messy to report and the radiation theorist's trials that deserve recounting. The latter was the high road to quantum theory and so is more interesting today than the search for the formula, though the formula was rewarding and is still quite admirable.

5.7 Atomic Spectra and the End of Atomism

Why were Kirchhoff's followers, especially Planck, inarticulate about the motive behind their researches? Their explanation, their claim that the law is universal, is too general; so is the explanation that they were not clear about matters because the situation was unclear. Clearly Kirchhoff was attempting to eliminate from his presentation as many details of spectroscopy as possible as he was studying the fact that emission and absorption are the negative of each other; it was clear to him that the correlation of the two should eliminate many details of specific mechanisms, of specific elements, since the reciprocity of emission and absorption holds for all elements.

But why was this so important? Taking science to be certitude, as Kirchhoff very much did, he could not possibly see that his very project was a mutiny against classical atomism. But in retrospect at least that is what it clearly was. In retrospect the initial situation was obviously problematic; trouble lurked in the very theory of valency, which is the core of Dalton's atomism that has rendered it the first scientific version of an ancient metaphysical doctrine.

Consider the question, what mechanism grants an atom valency? This invites a search for some atomic model, and a model describes parts of an atom contrary to the atom's image as a point particle or a smooth billiard ball. Does the atom have parts and if so are they separable? The discovery of the electron in 1895 suggested that the answer is yes. The Rutherford-Bohr theory says so explicitly. But even before that it was empirically established that radioactive matter disintegrates; and even earlier the suspicion was that the electricity of an atom is distinct from the atom itself, and so the charge, if it is associated with matter, is sub-atomic. Thomson's discovery of the electron was surprising as the electron is a charge associated with a particle of matter much smaller than the smallest atom.

This is presented not as hindsight but as a strong suspicion entertained at the time; the atomists Faraday and Maxwell denied the existence of material atoms of electricity. They knew that charge is discrete: Faraday established that to his own satisfaction when he linked charge to valency in his theory of electrolysis of the 1830's. Nevertheless they were reluctant to view electricity as atomic. For the idea that the charge is of a part of the material atom plus the idea that charge is detachable amounts to smashing the atom. The question is not, whether the atom can be dismantled, as we know that it can be. Rather, the question is, was there a serious suspicion that this is so? This suspicion, I propose, existed, and is the cause of the excitement and of the inability of such articulate people as Kirchhoff and Planck to explain the import of Kirchhoff's law. If anyone could be articulate about this matter it was Einstein. By the time he entered the scene, however, the atom already had decayed and he never doubted the divisibility of atoms. He took it for granted that electrons are bound to atoms by a fixed energy and he showed that the photoelectric effect can help evaluate it with the aid of Planck's formula. (More on this later.) The breakthrough was due to Rutherford, the inventor of the idea of sub-atomic particles. At the time it was known that radioactive matter radiates three kind of rays, alpha, beta, and gamma. Rutherford identified them: alpha rays are helium atoms; beta rays are electrons; and gamma rays are x-rays. Alpha and beta rays were known to be electrically charged, positively and negatively. The idea that the atom is a compound of a negatively charged electron and a positively charged nucleus was the most obvious explanation of the electric neutrality of atoms. Yet this was presented by Rutherford and Bohr only after the refutation of the theory which presented electricity as distributed in indivisible atoms. The originator of that theory was none other than J.J. Thomson.

The idea that there are sub-atomic particles, thus, preceded the development of the idea of quantum theory, of viewing atomic spectra as tools for the study of the structures of atoms. But this is more-or-less an accident: were there no radioactive matter easily available, or were the discussion of Kirchhoff's law as the study of atomic mechanism more explicit, then the idea of using atomic spectra as means for the study of the inner structure of the atom would have come first. Niels Bohr said he had developed his theory in order to account for the conduct of the Rutherfordian sub-atomic particles and only afterwards, when he asked himself, where else could the theory be applied so that it could be tested, he thought of atomic spectra. He said this was not a difficult idea to come by. Today Bohr's theory is presented as chiefly a theory of atomic spectra. Though this is not quite in accord with the historical development, it is not contrary to the general situation of the time, as then the search for the mechanism for spectral lines was well-known. Bohr's innovation was that the vibrating electron does not radiate unless it can lose relatively large bursts of energy; these were his revolutionary quantum jumps.

Chapter 6

THE BACKGROUND TO QUANTUM THEORY

6.1 The Stefan-Boltzmann Law

I do not know how to proceed from here, and I have given serious thought to the possibility that I leave matters at that and call it a day. And so, if you now call it a day and discontinue reading, I will not take it amiss. The point is that the story from now on requires more thermodynamics, statistical mechanics, electrodynamics and calculus than I am willing to take for granted. So, from now on all I can offer is a gross sketch, an outline with not enough detail for critical reading.

If your mathematical training is above the average you may wish to consult Appendix A. In any case, the mathematics directly involved in the whole of the rest of the story is no more than a page or two. I can also put formulas in footnotes, but I find this procedure unsatisfactory. For the formulas which set up these exercises are high-powered results of the diverse theories mentioned. Indeed, in some very significant cases the results were worked out, with sweat and blood, by the major participants in the story. A researcher may be fortunate and find in a colleague's paper, even in a standard textbook, just the formula needed for the specific purpose at hand. More often, however, the researcher has to work it out alone to suit that specific purpose. This is another way in which a smooth logical sketch distorts very grossly the actual events. Often enough the final presentation conceals ever so many faulty attempts, reasonable attempts that had failed, etc. But enough of that.

The detailed stories of the last stages of radiation theory—Boltzmann, Wien, Planck, Einstein—is offered in a number of books, such as Whittaker's *History of Theories of the Aether and Electricity*, or Martin Klein's detailed studies, or other works, including my own papers, mentioned in the *Preface* above. If you can read the calculus, you will find little difficulty reading these works. I really find little merit in repeating my own version of the story a second time. And were I to do a thorough job of questioning all the assumptions that make the deduction so quick and easy, then I should have to offer you a study of the history of electromagnetism, thermodynamics, and statistical mechanics. I cannot do so now.

What, then, do I offer in the rest of this study? I promise to try to keep you amused and not to try to convince you of anything; I will relate why I, for one, do not like the way science is usually described, and use the present story to illustrate this. In my opinion science is just a perennial mess and a constant source of puzzlement. This makes science less than admirable for some people; it makes it human for them, all too human. Now, the expression "all too human" is one of the nastiest witticisms of Friedrich Nietzsche, converting as it does the most laudatory epithet into a fairly pejorative one. Science is, indeed, all too human in just his sense, yet all the same, all the more admirable.

This is no protest of loyalty to science, and what stuffed-shirt professors think of this is of no interest. Professors in charge of the public relations of science are a mild public liability and no help to the advancement of science. Surprisingly many good scientific researchers covet their good opinions too earnestly. Even searchers for the truth at times seek the high opinions of the stuffed-shirt professors in charge of the public relations of science. For example, Ludwig Boltzmann, the hero of the present section, committed suicide partly because he was that kind of person (some people will not commit suicide no matter what), and partly, among other unhappy things, as he felt lonely and miserable and unrecognized by the scientific establishment of his day.

I do not wish to be pious about Boltzmann's suicide; most suicides are pathological, and for all that is known, so was his. On top of that, he had the wrong notion about recognition: when Max Planck, whom we shall soon meet, expressed disagreement with him, he took it as a sign of no recognition rather than of recognition. This is neither fair nor wise. It is particularly unfair because Planck praised him too. Boltzmann is the person who presented entropy as disorder and Planck said in the preface to his *Treatise on Thermodynamics*, I have just heard of it and it is marvellous and I am going to study it. And it is not wise because critical attention is attention and so appreciation. Indeed, Planck himself finally—in 1900—came to use Boltzmann's ideas, and from then on, Planck reports in his scientific autobiography, Boltzmann became friendlier to him.

And so, Boltzmann was an unusually sensitive soul in much need of acknowledgement and encouragement and recognition and support. This is why he, rather than, say, Planck, took the lack of recognition so badly. Nonetheless, as he needed it and as he had the right to expect it, he should have been granted more of it. So much for the human side of the story: like most world-shaking developments it took place not in the limelight but in relative obscurity—for example, the obscurity of scientific papers which are seldom read and scarcely appreciated yet which nevertheless change the course of events most significantly.

Boltzmann's idea is simple. The item sought, we remember, is the distribution of radiation in a cavity in a black body at different wavelengths and temperatures. For, you remember, in equilibrium the body (black or not) absorbs as much as it emits and so the distribution of its emission equals that of its absorption. Thus Kirchhoff's function $k(\lambda, T)$ is the same as the distribution of radiation in a cavity. Yet, before finding the answer to this formidable question about the distribution of radiation in the cavity for the different wavelengths and for different temperatures, we can ask, what is the total radiation in the cavity for all the wavelengths for different temperatures? (It makes no sense to ask what is the total radiation for all temperatures as the cavity does not have different temperatures at the same time; but it does radiate different wavelengths at the same time. So the sum of energies it radiates at all temperatures can be infinity as it is not a real quantity, but the sum of energies a body radiates in a given temperature over all possible wavelengths must be finite as it cannot possibly radiate infinite energy at a finite span of time; similarly, it cannot absorb an infinite amount of energy, and so the cavity cannot possibly contain an infinite amount of energy.)

The total quantity of radiation in a cavity is the total energy there, that is to say, the total energy of electromagnetic radiation there. What then is this total electromagnetic energy? Here Boltzmann used Maxwell's formula for the pressure of light. It should be noted that today no one is puzzled at the use of Maxwell's theory—it is quite respectable. Yet years afterwards Boltzmann wrote a letter to the British journal *Nature*, in which he explained why most of his own colleagues on the Continent were hostile to Maxwell's theory.

Boltzmann assumed that the cavity is full of gas at the pressure which the radiation in it has to be according to Maxwell and at the temperature of the cavity's walls. Here, for the first time, radiation is treated as a gas whose particles move about and whose energy is the cause of its temperature and pressure. This supposition defies the wave theory; it is distinctive of quantum theory. Personally I would have little hesitation declaring Ludwig Boltzmann the father of quantum theory. Of course, he would not know what to do with quanta: he treated light as waves so as to compute its pressure and then switched to viewing it as particles so as to compute its thermodynamic characteristics. This is standard practice even today, sanctified by a principle of Niels Bohr of 1927 or 1928 called the principle of complementarity. Boltzmann said nothing about the presumed particles of light which he (erroneously) subjected to the laws of classical mechanics and classical statistics. So did Planck. As we shall see, the first to deviate from this was young Einstein. And so perhaps I was rash—not Boltzmann but Einstein should be labelled the father of quantum theory, except that Einstein has already too many honorific epithets. But then the matter is of little import, and let us not belittle Planck.

Boltzmann calculated the total amount of energy in the cavity by pretending that the radiation in the cavity is a gas, taking its temperature to be that of the cavity's walls (since the system is in thermal equilibrium) and its pressure to be that of its total electromagnetic energy. This should raise the question, what kind of energy is radiation? The question was barely noticed; the idea was considered a mere analogy, and it was endorsed on the ground that it agreed with Stefan's empirical findings.

Stefan's experiment was questionable, and Boltzmann supported it by an analogy, not to say by questionable suppositions. But as it happens, the law was empirically confirmed as near enough to the truth, as far as we know. It is still employed empirically in day-to-day laboratory work. Its success, then requires a better explanation than Boltzmann's analogy. The explanation which was given next is, it follows from Wien's law. What is this law? We are coming to that.

6.2 Wien's Law

Willy Wien (he was publicly known by his nickname) did not look at radiation as a gas or as anything else. He only took Maxwell's result that radiation exerts pressure. The radiation in a cavity presses the walls of the cavity just as any gas would though, of course, very much less so. Measuring the pressure of a gas is much easier than measuring the pressure of light; the latter requires extraordinarily sensitive instruments. (Wien's work delightfully illustrates the difference between barren instrumentalism and robust opportunism, to use Einstein's idiom.)

Wien pushed to the utmost the idea that when factors examined are internal to a body that body can be placed in any surrounding convenient for the purpose of the examination: the result is universally valid anyway, and for obvious reasons.

I should not have said "obvious": in the Nobel lecture that Wien gave in 1911 he explained matters at great length. First he got rid of the black cavity and replaced it with a white one: he imagined a white cavity with radiation inside it, the distribution of which is the same as that of the radiation present in a black body. The formula for a perfectly white body is $a \equiv 0$, to be read as a is identically zero where, to repeat, a is the absorption coefficient of the perfect mirror and its being identical to zero means that for all wavelengths and for all temperatures it is zero. The formula for a perfectly black body is $a \equiv 1$, to be read as a is identically maximal, it is the absorption coefficient of the perfectly black body, and its being identical to unity means that for all wavelengths and for all temperatures absorption is maximal.

The white cavity is inessential, as Boltzmann's calculations show. Nor does it matter how is it possible to transfer radiation from a black cavity to a white one. Radiation is no fluid, of course; it runs away at the greatest speed possible. And just as there is no black body and no white body, there is no such transfer to perform: it is all in the mind, and in no way does it put constraints on the thought experiment. Nonetheless, to please some unimaginative scientists, let me describe how this is done: put a black body in a thermally isolated white cavity, keep it there for a sufficiently long time (in practice a very short duration will do) and take it out of the cavity with minimal disturbance of the situation. Or have a cavity with one black wall and a white screen to slide and cover it. Of course, all these descriptions are only means of making a bit more plausible a story which essentially may be as implausible as one wishes. What is compelling is not the experiment but the logic behind it.

How important is it to visualize a logical consideration? This is a broad problem. Geometricians draw pictures now as Euclid drew them then. The pictures should illustrate the logic, and the logic should stand on its own with no regard for the illustrations. (Good teachers draw their illustrations as badly as their students can take them.) Time and again an axiom is smuggled into considerations *via* the picture. Wishing to state the axiom explicitly, some geometricians prefer to do without pictures whenever possible. But we all compromise and we all pay the price of compromise.

Back to Wien. Consider a white cavity and a black body radiation in it, i.e., radiation with a black body distribution. Consider, next, the act of reducing the size of the cavity. This is an investment of work which will, of course, turn to heat or to radiation (i.e., electromagnetic) energy. Query: will the distribution of the energy in the cavity remain after compression as it was before, i.e., as in a black body?

The significance of this question is this. Kirchhoff's function k should hopefully yield the relation between temperatures and wavelengths of light, it should unfold different distributions of wavelengths at given temperatures. Now suppose we know the distribution of wavelengths for a given temperature, can we conclude from it what will be the distribution of wavelengths for another temperature? Suppose compression will, indeed, retain the black body character of the distribution. And, we remember, there will be more energy in the cavity after it is compressed and so it will correspond to the radiation of a hotter black body!

What Wien showed is both that the radiation remains distributed as in a black body and that there is a simple way to compute its change. The change of the distribution of energy in a cavity from one temperature to another is called displacement; the law of this change is Wien's law of displacement.

What has to be explained first is, very simply, how can the more compressed cavity have a different distribution from its earlier less compressed self? The answer is very ingenious: by Doppler's effect. The wave hitting a moving mirror is like a wave coming from a moving source and so it changes its wavelength. Now we can effect the change in two totally different ways and get, perhaps, two different distributions. First, take a cavity and use a given amount of energy to compress it. Second, take another cavity where size and quantity of energy are from the start the same as those of the first one *after* compression, and use the same amount of energy to heat the black body temporarily put into it, so as to alter its radiation distribution.

Question: are the distributions in these two cavities identical? Suppose the answer is no. Then, the entropy law, the second law of thermodynamics, says that when connected, the two cavities will tend to mix their radiation so as to equalize their distributions of radiation. Hence this traffic can be exploited to extract work out of bodies at the same "temperature"! That is to say, the amounts of work invested in the two similar systems are the same, yet the dissimilar results enable us to cause change and gain work, which means that there is no equilibrium between two similar injections of work to two similar parts of the system! This is absurd. Hence, their distributions must be the same: compressing a white cavity where radiation is distributed as in a black body only displaces the radiation to another distribution as in a black body.

It may be objected that the two systems are not quite similar. But they are. Take a single system in equilibrium and divide it by a partition; no work can be gained by the flow of energy from one part to the other. Now condense one part and heat the other by a black body so as to have the same temperature on both sides. If the distribution of radiation over wavelengths is not exactly the same in both, then for some wavelength they differ, and work can be gained there. Hence the distribution of radiation by wavelengths is the same on both sides. Given an alteration of the temperature in a cavity, Wien's law of displacement correlates the new and the old distributions of radiation in it. There is no point in presenting details: they are the empirical Stefan-Boltzmann law, Boltzmann's calculations and Wien's calculations of how wavelengths alter through reflection from moving mirrors; I do not know if they stand up to criticism. The historical significance of Wien's law is that, being true it helped Planck further his research.

The law narrows down the choice of formula for k:

$$k = k(\lambda, T) = F(\lambda \times T) / G(\lambda):$$

it depended on a function of $\lambda \times T$ divided by a function of λ. It follows that for any given wavelength λ, the distribution of radiation in the cavity displaces so that the product $\lambda \times T$ is constant where T is the absolute temperature in question (the absolute temperature of a body is its temperature Celsius plus 273) and λ is the body's maximal wavelength, which is more-or-less its visible color; hence, the hotter the black body, the more its visible color shifts from red to violet and further.

This law is, again, quantum-theoretical as it says that in order to shift light from red to violet we need ever increasing heat; but why should blue or violet be hotter than green or red? Quantum theory tells us that light is emitted in pellets, and the energy of each pellet is proportional to its frequency. It stands to reason that when the source of energy is thermal we need a higher concentration of energy to shoot out a more energetic pellet. But all this is hindsight. It is hard to say how Wien and others felt about this oddity; for the first time a great difference was found between the red and the violet, not only in vibration, but also in concentration of energy. Boltzmann and Planck were both very critically minded thinkers, yet they did not notice anything strange here. In his Nobel lecture of 1918 Planck reflected ruefully on the not so distant past, saying, "I was filled at that time with what would be thought today naively charming and agreeable expectations" that classical physics could help explain all this!

I cannot explain all this. The facts seemed strange and disturbing, of course: here lay the challenge. Yet time and again a seemingly stray fact can be brought to the fold. Soon after Wien discovered his displacement law (1893) Pieter Zeeman discovered (1896) that when a flame is placed in a strong magnetic field its spectral lines split into triplets, each of the triplets polarized differently (two polarizations perpendicular to each other and a circularly polarized one). In quantum theory this fact, the Zeeman effect it is called, played a very significant role in the transition from the old quantum theory (1913) to the new (1926); yet at the end of the last century it played an exactly opposite role; as soon as Zeeman's discovery was made, Lorentz found a classical explanation for it. Lorentz never gave up hope of modifying all 20th century physics to fit the 19th century conception of physics; Planck and Einstein soon developed similar attitudes. I do not comprehend their tenacity. It is often explained as dogmatism, yet a slight familiarity with their writing reveals how open-minded they usually were. Einstein's obituaries on his colleagues often presented them as extremely fair-minded and open-minded, and he himself was unbelievably free of prejudice. And so I simply fail to comprehend their tenacity.

Perhaps here is the place to mention that the difference between colors was not confined to Wien's law. The photoelectric effect that Hertz discovered already in 1887 showed that electricity is ejected out of a piece of metal when it is radiated with violet light, and more so with ultraviolet light but not when it is radiated with red light. In

1899 Philipp Lenard showed that the more intense the light, the more intense the photoelectric current it produced, but the intensity increases with the increase of emitted electrons, not with any increase of speed of single electrons. Again, quantum theory (Einstein in 1905 to be precise) presented light as pellets with energies larger when violet than when red, but at the time no one suspected it. The explanation of the fact that some wavelengths were preferred was based on classical resonance theory that lays responsibility for the preference on the natural frequency of the emitting or resonating body, not on any general preference which Nature has for red over violet as quantum theory would later have it in accord with the law that red pellets carry less energy than violet. The electromagnetic theory of both resonance and dispersion were tried in the hope that together they would restore the classical explanation of the choice of red for small energies and of violet for big. It was all to no avail.

Yet Wien used successfully some results of classical electrodynamics. Was it luck? Was it the result of many trials, most of which led nowhere? Yes to both questions. The success of radiation theory is puzzling. Not that it was stupendous—indeed, the failure to produce the expected result is what mattered most. Nevertheless, as in a modern detective novel, good investigators know that clues may be blind alleys and traps, but they do what must be done.

Wien had exhausted the general possibility to his satisfaction. He had used thermodynamics most ingeniously and all the electrodynamics at his disposal, and the Stefan-Boltzmann law. He narrowed down the choice of formula for k, we remember, to dependence on a function of $\lambda \times T$ divided by a function of λ. Planck tried but could not determine the function of $\lambda \times T$ by general considerations.

Here he took recourse to statistical mechanics as Boltzmann did before him. He did so reluctantly and did not do so until he had no choice, since the continuous radiation is not given to analysis meant for discrete particles. But, again, with the energy and pressure of radiation given, an analogy is possible; radiation may be taken as a set of particles, and so the distribution of radiation among the diverse wavelengths given at a fixed temperature may be computed.

All that is needed to make Wien's law quantum theoretical was to push the analogy all the way and say that light does indeed consist of particles, and that the energy of each of these particles is proportional to the frequency of the radiation in question. It took two people to make this daring step, Planck and Einstein. They could read this off of Wien's distribution formula. But this was both too early and insufficient. Too early since only despair led Planck to feel free in 1900 to introduce specific statistical hypotheses, and insufficient because only in 1905 did Einstein show that viewing radiation as corpuscular was not mere analogy.

6.3 The Red Herring of the Violet Catastrophe

There is a red herring here, a chapter in the story that does not belong here at all, the so-called Rayleigh-Jeans theory which Sir James Jeans has labelled the violet catastrophe. Martin Klein, and since then others, noted that Planck was unaware of Rayleigh's work which appeared in 1900 after Planck had published his law but before Planck read his paper on December 14th, 1900, the official birthday of quantum theory. Now, when some claim for priority for a law is made there is no need to ask, was someone else—anyone else really—aware of its existence? All that is technically required is the date on which a law was made public. How is such a day determined? It used to be the publication date in a learned periodical. After a case of priority dispute (due to the fact that a paper based on a lecture was published before the lecture itself was published) in 1831, it was recognized that a lecture before a scientific body is also a means of publication. This is not the end of the story, and priority disputes may, indeed, depend on the question of publication date. Yet in Rayleigh's case surely his law was published in a scientific journal. Does it matter that Planck knew about it or not?

One answer is very simple: Rayleigh's law is false; it is only approximately accurate for low frequencies. Now this answer is very unsatisfactory. Many laws have turned out to be mere approximations to better laws, and this can be very problematic, yet many of them are well entrenched in our tradition: in particular, Wien's law (which, we shall soon see, was also modified by Planck as it is only approximately accurate for high frequencies; inaccuracies are not so important, especially in the exploratory stage—but then science is constantly in the exploratory stage).

The trouble with Rayleigh's law is too obvious; he always knew that it cannot be true. To prevent all misunderstanding let me add at once that he wisely and clearly said so and even made a big point of it.

His big point is very simple and so is the error. The function k was found by Wien to be the product of a simple function of λ and a function of $\lambda \times T$. He took the function of $\lambda \times T$ to be a statistical function, namely, one whose sum over all possible wavelengths is 1, as all probabilities or all statistical functions have to do (the sum of the probabilities of all possible outcomes of a given experiment should be unity). He also took the function of $\lambda \times T$ to be a product of $\lambda \times T$ and a constant. Now, as wavelengths go from zero to infinity, the sum of k over all wavelengths will be infinity (see below). This was the error which he pointed out as a serious fault.

An infinite k means that there is an infinitely large quantity of radiation energy in any cavity. It also means that e/a is infinite, which is silly as it says that, given an empty cavity and radiating walls, it takes an infinitely long time to create an equilibrium, which means that the walls will radiate and despite their absorption they will always cool down by radiating until they are absolutely cold, that every piece of matter will always lose thermal energy to the surrounding space.

It is particularly amazing, to complicate the story further, to have Jeans associated with Rayleigh here. He came late, in 1905, to a story that happened in 1900, suggesting that perhaps the leak is very slow so that the trouble is tolerable. At best this is dodging the empirical refutations of Rayleigh's law, not the idea that equilibrium is *a priori* impossible: it shows that this famous fellow did not see the situation correctly, that he barely understood it.

Now Rayleigh knew that the law is false, and said so. Why, then, did he publish it? The answer is that he deduced his false law from an accepted important principle of statistical mechanics thereby suspending that principle. In other words, he created, or rather discovered, an important problem.

It is significant that the point at issue is the creation or discovery of a problem, since the deduction itself is not quite new (it was a variant of one previously offered by Maxwell). Rayleigh suspended the principle of equi-partition of energy. It is this: the energy of a system goes equally to all degrees of freedom so-called. A degree of freedom is a fancy name for an independent coordinate which offers a way to invest energy. Thus, a single billiard ball moving in space has three degrees of freedom to begin with, the three dimensions of space, whereas moving on a billiard table it loses one degree of freedom as it can move only in two dimensions by rolling. The rolling itself adds nothing to the degrees of freedom, since the rolling is determined by the speed of the ball and by friction; it is not a degree of freedom. A billiard ball in space, however, has the three degrees of freedom of the three spatial directions and one or two (figure which if you can; the answer is two) for rotation. A point atom has three degrees of freedom and two atoms have six, but if they are tied, they have only four or five (depending on whether they are tied together rigidly or elastically). Now Maxwell was worried about the billiard ball. If it is absolutely rigid, it has five degrees of freedom. Otherwise, if it can vibrate elastically it has infinitely many frequencies, each of which is a different degree of freedom. By the principle of equi-partition of energy every finite energy should be divided to infinitely many degrees of freedom, with zero as the outcome. Maxwell did not like the elastic billiard ball model for the atom, and he did not like the absolutely rigid model either: at times he preferred point atoms.

The problem is not the one created by Maxwell but by Rayleigh; he found something amiss in statistical mechanics. Let me address this problem.

There is an aspect to the problem which is conservative, and which I therefore rather dislike: there is something faulty in radiation theory, and hence at least one of the ingredients that go into it must be faulty. The three ingredients are electrodynamics, thermodynamics, and statistical mechanics. We can rush to blame the newcomer and save the two established doctrines. It is a fact that Rayleigh was unable to follow Planck, let alone appreciate him, and so the conservative element is present here perhaps not by accident.

Yet Rayleigh was serious about statistical mechanics, and was as right about it as Maxwell. Quantum mechanics views the two difficulties, Maxwell's and Rayleigh's, as one, and solves it by rejecting the theory of equi-partition of energy. Quite inadvertently Bohr's theory of the atom solves the problem for Maxwell intuitively; his theory gives a rule of distribution of energies for atomic vibrations; and quantum statistics solves Rayleigh's problem by offering a substitute for the law of equi-partition. Hence, whatever I say about the problem introduced by Rayleigh, as a red herring, I would like to stress that it was a respectable problem elsewhere; it is not my aim to debunk it.

The problem caused by the infinity of the degrees of freedom of the cavity did not trouble Wien or Planck when they applied statistical mechanics to radiation, as they intuitively evaded it. Ever so many explorers evade ever so many problems; naturally, too many problems paralyze. This is why we seldom have priority for the discovery of a good problem (there are notable exceptions, of course). Yet, just as too many problems paralyze, so too few and poor problems create a lot of mediocre research results, not to mention the less than mediocre. The whole study of questions and how they are discovered, classified, and rated, is an innovation that I cannot discuss here. Let me only mention again Imre Lakatos, who has observed that a wonderful problem may degenerate as a result of an over defensive attitude or of some change of scenery.

Many writers, including Wien himself, prefer to present Rayleigh's law not as a problem but as a conjecture. Rayleigh's law fits long waves, Wien's law fits short waves, but Planck's fits all. This is a nice piece of fictional history which was probably invented by Jeans, so as to let him have a piece of the action, but soon enough it was endorsed by Einstein, who was at the time at pains to illustrate in any way he could the idea that scientific theories are (not perfect but) series of approximations, each approximating its successor under certain well delineated circumstances.

(This presentation of the Einstein-Popper view of the dynamics of scientific progress is somewhat uncritical; I confess I do have some misgivings about it. Yet the purpose of the present study is to present this view as a superior tool for the study of the history of science; I have criticized that theory elsewhere, but here I must leave it.)

6.4 Planck and Bohr on Models

Literary devices are not suitable for histories of science. In this part of the story the pace quickens and the climax is at hand; yet I cannot quicken the pace of my narrative.

To begin with the chronology of the quickening pace: the bits and pieces collected since 1800 or so started making rudimentary sense with Leslie and Prevost around 1810 or 1820 with researches that culminate with Kirchhoff's law of about 1860. The pace quickened, but new results got harder to come by though the number of clues mounted, significant errors in the clues got cleared, and some preliminary order was put into them. Of all attempts to make sense of Kirchhoff's puzzling and exciting law, we can report only the Stefan-Boltzmann law of the 1880's and the Wien law of the 1890's. To be precise, Wien's important papers are from 1893 and 1896, the latter being the year which sees Planck entering the picture. The photo-finish, you remember, is December 14th, 1900.

So much for the quickening pace. In a well-written action novel all material preparatory for the comprehension of what goes on at an ever increasing speed must be laid down clearly during the early parts of the story. The capable author prepares good sub-plots each of which is designed to familiarize the reader with another important item that will be used quite unhesitatingly in the zestful grand finale, and all the author has to see to is that the sub-plots fit the story line tolerably well.

There is nothing like this in a history of science. The grand finale has a new hero, and whoever has heard of a new hero introduced so very late in the day, utterly unannounced? But such is the story of all true adventure, scientific adventure included. The entry of Planck is marked with a very strong idiosyncrasy: a new idea that was not clear enough even to Planck, much less understood by his contemporaries. And, the worst of it, Planck was no *avant-garde*; rather, his leanings always were on the conservative side. He was never quite a stuffed shirt though; he was the one to discover young Einstein with whom he struck a lifelong friendship. His conservative views, in science as well as in politics, never bothered Einstein because he was willing to argue openly and fairly. His new idea which led his early researches was in effect a bridge between the old and the new; he was a rebel not welcomed by either conservatives or revolutionaries. Let describe his chief idea, even though it is quite philosophical, and doing so will spoil the quickening tempo of the dramatic narrative.

The topic at hand is models. Many interested non-physicists have had the occasion to hear Niels Bohr's dictum: thou shallt not make models. They must also have heard of Bohr's own model, the thing (what thing? how thingy?) that earned him his Nobel Prize. If they are more than cursorily interested, they may also have heard of models of the nucleus of the atom, such as the celebrated drop model (related to the nuclear bomb—a drop of water, when big enough can split into two or more, an idea which was tried on the nucleus soon after the first observed nuclear fission). This must sound confusing. It confused me in my student days and it made me undertake the studies whose findings I am now describing. I found that my teachers and most physicists I met personally were equally confused—until I heard the philosopher Karl Popper narrate the background to the story. So now, instead of the promised photo-finish, we have a long, interesting flash-back which should clarify matters. Sorry.

The idea of models is central to Plato's philosophy which played a crucial role in the rise of modern science as we are told by the greatest historian of science of our century, Alexandre Koyré. According to Plato what makes a thing what it is, is its shape—its geometry (meaning stereometry). This cannot be because not three but four dimensions have to be considered: a fast moving body differs in so many respects from a slow one! Even Plato offered a fundamental law of nature which relates not to space but to time; all things, he said, deteriorate, lose their shape: a coin fresh from the mint is in its best shape, but time nibbles into it and finally destroys it.

No matter. The leading Renaissance researchers were Platonists as William Whewell noticed already in the last century and as the no lesser historian of science, Alexandre Koyré, showed in great detail and in an exciting manner in our own century. The Platonism of the Renaissance led to two decisive characteristics of classical physics, one is the neglect of time and the other is the neglect of scale.

Let us take first the fact that time was neglected. Processes were described as series of spatial configurations; for example, planets move in orbits. After Einstein everything has its world-line, its four dimensional line. Also, time order was deemed unimportant: processes were deemed reversible. This, true enough, clashes with Plato's law of decay, as noted by R.F. Jones (*Ancients and Moderns*) in this century and by others. But then Plato's law of decay clashes with Plato's own geometrism, or at least the one is a mere rider on the other. Moreover, intellectually speaking, the Renaissance was possible only after the law of decay was repealed.

I do not wish to be drawn to the question whether the Renaissance was an economic, a political, or an ideological matter. Nor do I deny that ideology can be inconsistent. I only wish to repeat the known observation (on which there is a vast literature, including Jones's book) that some time before 1500 a process began that ended in the scientific revolution, during which scholars, collectively and imprecisely known as the Renaissance thinkers, were aware of the business at hand and consciously rejected the law of decay.

You may remember that the law of decay has had the last laugh, that today we call it the law of entropy or the death of the universe. It was Maxwell who noted that something unclassical, a revolution in fact, had entered physics through the back door. He proved this with a thought experiment that is now a myth in science and I am glad to explain it properly.

Imagine a demon—the celebrated Maxwell's demon—sitting near a flood-gate on a dividing screen between two vessels filled with a gas. The floodgate is open; equilibrium is achieved and both sides have equal temperatures. Now the demon closes the gate and watches. When a molecule on the left approaches the floodgate, he lets it pass if it is rapid, and he stops it if it is slow; when a molecule on the right approaches, he stops it if it is rapid, and lets pass if it is slow. More precisely, the floodgate is bombarded with molecules hitting it from both sides, and the trick is to have the floodgate open when there is more energy flow from left to right and closed when there is more energy flow from right to left. This is a lengthy and tedious business since most molecules have average velocities and so most of the time the floodgate is bombarded on both of its sides with molecules carrying the same amount of energy. But fluctuations must occur, however rarely, and the demon, child of Maxwell's imagination, has all the time in the world. He thus creates a disequilibrium out of equilibrium at no cost contrary to the second law of thermodynamics. The demon can do better: he can let more molecules pass from left to right than the other way. This will create an even larger disequilibrium!

Why was this demon invented and what is he supposed to do? He is supposed to make us ask: is he possible? Of course, he is possible logically but not technologically. But, if you have fared with me thus far you know that thought experiments should elucidate theories. Try this. Suppose we knew nothing about the universe except that it obeys Newtonian mechanics; is then the demon permitted entry to the universe? Yes; Newtonian mechanics permits Maxwell's demon entry to the universe. Suppose we add thermodynamics to Newtonian mechanics, is the demon still allowed entry? No; Maxwell's demon is most clearly deported by the law of entropy. One may still ask: is there perhaps also a mechanical reason for his exclusion? Can the energy demanded for the moving of the floodgate be so much as to compensate his upset of the equilibrium? (You remember that refrigerators upset the equilibrium but demand enough expenditure of energy to handsomely compensate the law of entropy.) This excuse will not do: there is no law of mechanics that imposes a minimum mass on the floodgate or a minimum expenditure of energy on the operation on it. All studies of this kind regularly assume that the energy required for operating a valve is practically nil, so that in energy calculations it can be neglected.

Hence, the entropy law outlaws what classical mechanics permits. Thermodynamics can never be made a part of mechanics as it has an additional postulate, Maxwell concluded reluctantly.

Maxwell presented a simpler proof for the independence of thermodynamics from mechanics. Classical mechanics is time reversible; the entropy law is not.

Suppose we decide to use a time coordinate which is the reverse of ours, i.e., if we put a negative number for future times and a positive number for past times, instead of the usual way, and if we agree to always count-down rather than count-up (and *vice versa*, of course). The change is only of a convention of marking time, not a change of any view of the world; it is but calling the past "future" and the future "past". All conservation laws are indifferent to this change, as they equate past and future characteristics of systems, and Newtonian mechanics may be viewed as a set of conservation laws. Yet this change will alter the entropy law. (Animated cartoons of processes which take place according to classical mechanics, may run forward or backward without exhibiting any violation of the laws of mechanics; but they will violate the entropy law: an animated picture of an explosion run backwards, for example, looks obviously wrong.) This again proves, says Maxwell, that the entropy law is not a part of classical mechanics but an added assumption.

So much for the entry of time into modern mechanics in a way that Platonizing Renaissance thinkers would find hard to take. The other result of Platonism, or of model theory, or of geometrism, is not so much inherent in Plato's philosophy as in the geometrical theory of his disciple (two generations removed), Euclid. In Euclidean geometry size plays no role; the choice of a unit of length is utterly immaterial there. It does not affect and is not affected by the rest of the system. And so the idea developed that a miniature model, if accurate enough, is as good as a replica.

This idea is of immense historical significance. The mystic cabalistic theory of Man as microcosm depends on it; its corollary was that Man has dignity and so is fit to perform scientific research; it played an enormous role in the philosophies of many thinkers from Giovanni Pico della Mirandola (before 1500) to Galileo (after 1600) and the latter was aware of the difficulties surrounding this idea. A striking example: a model-airplane can fly yet its large replica cannot. Also, small ships and large ships sail the ocean, but, other things being equal, large ships are more economical. (Formally, every non-linear function of length will violate the model theory.) Yet the question remains, is there a natural unit? is there a difference between magnitudes which enables one to speak of size not relatively but absolutely?

Yes. Planck and Einstein first brought about a physical theory that distinguishes absolutely between the large and the small! And the result that Planck gave rise to, quantum theory, had an immediate connection with the peculiarity of the entropy law, of the second law of thermodynamics, namely, its independence from geometry and its dependence on time, as I shall now explain. Before coming to that, let me explain a bit more about models.

A Platonic model follows the law of geometry. Yet when discussing Maxwell's idea, we asked, does entropy have a Newtonian explanation? Does it have a Newtonian mechanical model? Here we have different senses of the word "model".

The word "model" is nowadays used in diverse ways, some of which have little or nothing to do with physics, e.g., a model can be an individual exhibiting fashionable clothing, and a model, or a role model as it is called, may be an individual whose conduct is to be emulated. There can also be, in this very sense, a model of a scientific theory, one whose qualities, or some of them, should be present in our next theory. Thomas Kuhn, the philosopher and historian of science, has called the model theory a paradigm (in Greek, a chief example); in his opinion one cannot say what qualities make it a model to emulate.

A theory in science is expressed as a universal statement. To apply it to a particular case we must say something about that case, and what we say is a statement of initial conditions so-called. From the universal statement plus the initial conditions we conclude further statements about the particular case at hand, and these may be test statements, predictions or forecasts, specifications, etc. The initial conditions may be peculiar to one case at hand, or they may describe a whole set of cases; they are then a model. The initial conditions of one model in a fashion show will be copied by the thousands, hopes the fashion designer, and so the dress shown in the fashion show is a model just in this specific sense. If we have all the possible initial conditions characteristic of one hydrogen atom, then, surely, those are also characteristic of all hydrogen atoms. This is the case of Bohr's model of the hydrogen atom.

Models have a funny status, then. The initial conditions of a particular case proper are accepted without further ado but merely for particular purposes (the technical term here is *ad hoc*: the initial conditions are accepted *ad hoc*); a universal law, on the contrary, is accepted on its own merit yet there is the desire to see it integrated in a view of the nature of things. A model is somewhere in between. Where do models fit? They do not fit anywhere and should perhaps be banned or at least declared transient.

History ruled otherwise. The seventeenth century philosopher, mathematician and scientist Descartes was a significant role model for many for many generations. He recommended the use of hypotheses regarding mechanical model in physics and in biology. His recommendation legitimized the use of models and gave it class. His recommendation likewise legitimized the use of hypotheses and gave it class. This led to developments which have to be skipped as we plunge into the epoch of the model in the late nineteenth century. Its champions were the followers of Descartes, Kelvin and Maxwell. They assumed that electromagnetism was part and parcel of Newtonian mechanics, but they could not deduce the one from the other though on their assumptions this should have been possible. They (followed a tradition that began with Descartes and included such illustrious thinkers as Fresnel and Ampére, and) tried models, assumptions about mechanical properties of the ether, mechanical properties which should characterize the ether, arrangements of specific initial conditions, elasticity arranged this way, fluidity arranged that way—very much like a complex clockwork but with cogs, pulleys, springs, fluid circuits, etc., etc. No matter how complex their models were, they failed: they could not build a mechanism that would satisfy them, one that should nicely obey both Newtonian mechanics and electromagnetism.

Newtonianism was not the universally accepted theory that many think it was, though it is the view (the world-view or the picture of the world, or the philosophy of nature or the intellectual framework) that came nearest to being universally endorsed among research physicist and so it was the unofficial official doctrine of the scientific community (as politicians say). At that time the opposition, especially Faraday, maintained that forces could fill empty space and follow laws of their own. Also, at that time, around the middle of the century, when Helmholtz was still young and unknown, he wrote a revolutionary paper conservative at heart, *On the Conservation of Force*. The paper opens with an expression of faith in Newtonianism: an explanation is satisfactory, it says, only if it assumes the existence of atoms in a specific configuration obeying a specific force acting at a distance. But, he added, in the meantime, as a stop-gap measure, the law of conservation of force can be held. He and Tait and others had an excuse for employing the heretic theory of the conservation of force: since action equals reaction, the sum of all forces acting in a system is always zero, and hence forces are conserved. Also, since the law of conservation of energy (which was soon deliberately confused with the law of conservation of force) holds for all mechanical systems, it may be applied to systems whose mechanisms—or models—are still unknown. This view was also expressed by disciples of Faraday, such as J.J. Thomson who won the Nobel prize for his discovery of the electron.

Planck understood Helmholtz differently. He saw in the work of Helmholtz, his admired teacher, an attempt to elevate the law of conservation of energy to the rank of a principle, thereby rejecting the view that the law has the status of mere consequence of the equations of Newtonian mechanics for some known or unknown models. Planck named the law the first law of thermodynamics in order to stress its priority over any model. This was his intellectual framework, the guiding idea of his researches. Yet there is here a nuance on which he was historically inaccurate; Helmholtz preferred to have a model for electrodynamics, but in a pinch he was willing to make do with characteristics that a model must have, such as the law of conservation of energy. But Planck heard him say, I do not need the model, and I hardly want it. Perhaps Helmholtz started conservative and ended more revolutionary. Perhaps, however, Planck had a new idea and he pinned it on his teacher as it was too revolutionary for his conservative taste.

The second law of thermodynamics rendered matters more clear-cut as Maxwell proved that it can have no mechanical explanation and so no mechanical model. So Planck wanted to elevate this, too, to the level of a universal principle as no model could satisfy or interest him. Planck tried to present the two laws on a par, the first and second law of thermodynamics as he called them, ignoring the fact that the first law is a theorem in classical mechanics and the second is an added constraint. This ambiguity and halfheartedness was finished with Bohr who said, there is never a way back to classical physics or to any physics that provided every particle with a clear world-line. In this sense Bohr barred all models. Yet he had models all his life, quantum-theoretical ones, partial ones, that were not capable of satisfying the nineteenth century model theorists, Maxwell in particular, who wanted all physics to be parts of Newtonian mechanics.

I shall now explain why all this mattered so much to Planck: it made him act as he did though the end result turned out much different from what he expected and he never overcame the disappointment.

This is how Planck's aversion to all models and Bohr's aversion to mechanical models contributed to their seminal researches.

6.5   Planck's Law

I hope I do not have to explain to you why Kirchhoff's law fired Planck's imagination. Of course, Kirchhoff was his teacher as Helmholtz was. But his admiration for them is but a minor item. In his scientific autobiography he complained rather bitterly that they refused to appreciate his work. And evidently the frustration caused by his teachers went deep because all the honors he received from contemporaries and followers left him, as he portrays himself in his scientific autobiography, disappointed and frustrated to the end. But what matters much more is that Kirchhoff, as well as Boltzmann and Wien, found general and important applications of the law of entropy.

Planck approached his task religiously—his preparation was long and severe, and execution of his task absorbed long stretches of extremely long and hard work. He began by a very careful study of the law of entropy. In particular, he tried to view every system undergoing change as one which has a numerical function describing its entropy and to correlate the process with thermodynamics, mainly its two laws. And his main rule was, we remember, avoid any specific assumption about the system at hand—thou shallt make no model!

It is important to notice that Planck was the first to apply the law of entropy to the radiation in the black cavity. No doubt Wien already claimed to have done so: he said in 1897 that he had previously known that "the entropy of radiation of a known intensity and color can be determined", but what Wien did, as Boltzmann before him did, is work on an analogue. They both viewed the radiation enclosed in a cavity as if it were a gas. Plank presented the entropy of the radiation as a radiation leaving open the question, is it analogous to a gas? Or is it really a gas? He thus opened the way to the most incredible answer: yes, radiation is a gas!

I do not mean to report that radiation is, indeed, a gas. Quantum theory says it is a special sort of gas. I mean to report that Einstein seriously entertained that hypothesis in 1905. And I should draw your attention to the fact that this idea does not make sense unless approached from a relativistic point of view permitting radiation, which is pure energy, to be viewed as a gas, which is material. The traditional presentation of Einstein's work on relativity and on quantum mechanics, as if they were initially separate projects, strongly clashes with his view of himself as passionately concerned with metaphysics, that is, with an integrated picture of the world.

To back track. Planck hoped to calculate the energy and entropy, both in the walls of the cavity and in the space inside it, so as to see what distribution of radiation results in the case of equilibrium. He first assumed that there are charged particles oscillating in the walls of the cavity. These are called, understandably, electric oscillators. He stuck to the name of oscillators as he did not want to discuss the question, are they molecules, atoms, electrons, or anything else?

This point, again, characterizes Planck. What he tried to do is see how the existence of oscillators of various frequencies effects the distribution of frequencies in the cavity. For example, a wall that radiates one frequency, in equilibrium with the cavity, did it contribute to other frequencies? The answer is, no. Resonance only permits a resonator to absorb and emit the same wavelengths, its own. The black body radiation is distributed in a specific way; but why? Both Maxwell and Boltzmann approached distributions by assuming certain statistical hypotheses, certain laws of distribution. Not so Planck; he refused to make any hypothesis concerning any specific distribution as this too is but a model. In equilibrium the same distribution of frequencies in the cavity must be that in the walls; that should do. Planck found it easier to work with frequencies rather than with the traditional wavelengths as he studied the interaction between waves and oscillators, and the latter have frequencies but not wavelengths. Assuming thermal equilibrium between the walls and the radiation, he concluded that whatever the distribution of energy was, it is a simple proportion between oscillator and radiation.

The important point in Planck's work is his suspicion of any specific assumption and insistence on working only with most general ones. He later reluctantly accepted the theory of normal distribution of energy, but only to correlate the distribution of energy in the wall of the cavity to that of the cavity (in the case of equilibrium, of course). To begin with, he flatly rejected all other parts of statistical mechanics. He did not pay attention to Boltzmann's law of equi-partition of energy to all degrees of freedom and so, however near he came to Rayleigh's absurd formula, he was never in any risk of deducing it. To begin with, he rejected Boltzmann's view of entropy as a function of probability.

What Planck was looking for, in accord with this line of research in general, was a formula for the entropy of both the walls and the cavity which would tie together all he knew, namely that which correlated the two distributions of energies. He was very excited as he had success and he pinned great hopes on it. It was a false hope.

Here is an example with a moral somewhere. I have not given you enough evidence to inconvenience you. Had I shown how many false starts and uneasy moves each of Planck's predecessors had made, had I mentioned the few authors who worked in manners similar to Wien yet received somewhat different results, you might have wondered more. But at least, we can say, here is a clear and easy result. Here Planck supplements Wien's distribution law in a logical manner. In thermodynamics the entropy of a gas was introduced first and the statistical model followed suit; in Wien's work there was an analogy to the mechanical model, but not a complete working out of the thermodynamic equations. In Planck's work the analogy was pushed further, and it had a great success for no reason at all and to no good purpose. The snag was now empirical: Wien's distribution was empirically refuted just then.

In the old days of classical science refutations were deemed deadly. They were not always deadly and they were not always treated as such. The example of Kirchhoff's law illustrates that researchers were never as severe as they demanded of themselves. Einstein has rightly called them opportunists.

Planck was a staunch realist; though he was willing to set aside the refutations or seeming refutations of Kirchhoff's law, he was not willing to ignore the refutations of Wien's distribution. He had excellent reasons. As his *Treatise on Thermodynamics* amply shows, it is extremely difficult to avoid misstating or misapplying the second law of thermodynamics, the foundation of Kirchhoff's law. But here the case was clear-cut; indeed, the merit of Kirchhoff's law is that it helps make things clear-cut, so that refutation of it may lead to qualifications that we may not yet know and prefer to ignore for a while, whereas the refutation of Wien's distribution was decisive.

Perhaps I am in error to say that nothing came of the simple theory which describes the entropy of Wien's distribution. For, when Wien's distribution went overboard, Planck made a slight modification of his formula, though, it is true, he was helped by experiments, and derived from his new formula for entropy his celebrated formula for the distribution of radiation in a black cavity.

Planck's distribution is the first formula in the history of quantum theory; it has remained fixed and unmodified through the revolutions that the field has seen and it is endorsed as true in all text-books to this day. And so they all get excited about it even though, as we shall see, it silently dropped out of the picture in 1905 and came back only in 1924. The reason for the excitement is, of course, that Planck's constant, h, appears in his formula for the distribution of energy in the cavity. Quantum theory begins not with Planck's quantization of the radiation in the cavity to chunks or bundles, as these appear, you remember, in the study by Boltzmann; rather, the significance of Planck's quantization is that the light chunks or bundles are endowed each with the quantity of energy, E:

$$E = h \times \nu,$$

where h is an extremely small quantity so that the chopping off is fine enough to give a semblance of continuity. This formula is, of course, the one that explains the fact that violet light is more energetic than red as proved in the experiments of the photoelectric effect as shown by Lenard. The formula of the quantization of light is much more important than Planck's distribution for light in the cavity. Though the formula is from 1900, the constant h is from 1899. In a paper submitted for publication in 1899 and published early in 1900 Planck estimated the value of h, and in another paper submitted in March 1900—you see how productive he was!—he improved the estimate. Moreover, in the earlier of these papers he comes back to the Platonic idea that the laws of nature should include no reference to size, but with a twist: rather than have no magnitudes at all, we may have a theory using only natural magnitudes (such as the speed of light and the radius of the electron and the time it takes light to traverse it, to use Einstein's favorite examples)—this idea has a built-in distinction between big and small, of course. The reason Planck felt his constant to be so important is that it is a new atomic constant of an unusual dimension (action it is called, namely of the energy unit multiplied by the time unit; remember that the frequency of an oscillator is reciprocal to the period it takes for the oscillator to move to and fro). We still do not have enough constants to execute this grand program, but Planck never gave up hope, nor did Einstein, and Heisenberg and Pauli, the elder statesmen of quantum theory, got very excited in the fifties when they felt they came closer to it but they were, alas! on a wrong track.

So much exciting material was lost on the historians of science because it is more dramatic for quantum theory to have an official birthday. Even Planck himself in his very famous scientific autobiography leaves matters obscure.

And then Planck ran out of ideas. He wanted to know "the real physical meaning of this equation", his equation for the distribution of frequencies in the cavity which was in agreement with the facts. He worked hard and tried hard and was at his wits ends. Finally he capitulated and was willing to employ a model as an "act of desperation" and never wholeheartedly till the end of his life! He accepted not only statistical mechanics in general, but also Boltzmann's theory in particular, the whole of it. In order to apply Boltzmann's theory here, radiation was considered a gas proper—not compared to a gas, but actually quantized. This, too, was not terribly new. Space is continuous and velocities are continuous. Boltzmann divided space and velocities into discrete elements in order to count the number of all the possible distributions of atoms in these elements so as to calculate the probability of each but this division was only a part of his calculation, to be eliminated before the final result. Planck's formula came out of these calculations and his final formula is one in which the discontinuity is not eliminated. Had he applied this theory earlier in the day he might have easily obtained, already in 1899, the absurd Rayleigh distribution. But, because he had his distribution first with no recourse to Boltzmann's model and only later tried to fit his distribution and entropy via Boltzmann, he found that he could do it using his equation, $E = h \times \nu$. As he said over thirty years later, "this was a purely formal assumption and I really did not give it much thought except that, no matter at what cost, I must bring about a 'positive result'."

Before you jump for joy, I must tell you that the calculation was not quite right. But this is the story of Einstein.

6.6 Einstein and the Photoelectric Effect

I am overshooting my mark, having reached my target, the year 1900. So I shall be extremely brief and consider this and the next section mere mopping-up operations, tying up loose ends, a sort of epilogue, like in the novel when the mystery is solved but the author has to tell briefly what happened to the story's characters in the aftermath.

Einstein was a young patent tester fresh out of university (technical college, to be precise). He had a keen sense of significance and tried hard to ignore all detail if he could. He wrote in 1905 three explosive papers. They were almost his first shot and each of them was a bull's eye. One was on relativity. I shall repeat two relevant points about it. First, it replaced classical mechanics with a newer theory that allowed radiation to exist in the form of waves in empty space without a mechanical model: forces could exist and vibrate in the vacuum. Second, it equates matter with energy thus blurring the distinction between the two, thereby permitting one to view radiation as a gas. (Nuclear explosions do not enter our present story.) His second paper, on the Brownian motion or fluctuations, showed that statistical mechanics was not in so excellent an accord with the law of entropy as understood before statistical mechanics; for, as statistical mechanics describes random distributions of energy on the atomic level, on that level it allows fluctuations, deviations from the law of entropy. Now fluctuations were always part of statistics. They had already troubled Boltzmann as well as other atomists who all tried to dismiss this bother as negligible, on the ground that as atoms are small their fluctuations cannot be measured. This is the wrong attitude to scientific research: scientific research is the study of problems, and though researchers are at liberty to ignore some problems while concentrating on others, there is never any need to shoo them away. To shoo troubles away is to prefer that they do not exist in the first place, to prefer the state of affairs in which scientific research is uncalled for. Moreover, fluctuations are known to be possibly observable; Maxwell's demon observes them, and if we too could, then we could find counter-examples to the classical entropy law. Einstein was undeterred and showed that fluctuations can be seen and even help us calculate atomic magnitudes. Planck agreed with Einstein that this, indeed, has turned out to be the case, and he withdrew his own, by that time the best ever, wording of the second law of thermodynamics. Recently a few modern writers, notably Karl Popper and (his follower at the time) Paul Feyerabend, expressed agreement with Einstein. They were responding to other researchers, primarily the great quantum physicists Leo Szilard as well as Leon Brillouin and Dennis Gabor, the inventor of the hologram, who had led a successful quiet revolution. They had introduce a new factor into the considerations—the energy which the demon has to spend in order acquire the information he needs for the successful performance of his job of upsetting the second law of thermodynamics. They suggested that adding this to the calculations will rescue the law and thus restore the *status quo*. Why is the added restriction on the demon? after all, the demon is excluded by the second law of thermodynamics, and so there is no need to add any restriction on his conduct! The answer is simple: Einstein's study points at deviations from the entropy law; this permits one to emulate Maxwell's demon and circumvent the second law of thermodynamics. What is at issue here is not a matter of practical application: as Popper stresses, if the demon is possible, then the heat engine which he runs violates the entropy law, though the entropy law still holds well enough, as the demon's engine is the most inefficient heat engine possible—as long as we have heat sources to utilize. The attraction of the crazy arrangement of the demon is, of course, in the possibility to pump energy out of matter in ordinary temperatures: we will not try this as long as we have concentrated energy, but the question is, can it be done anyway? This question, raised first by Einstein, not by Maxwell, is still controversial, almost a century later.

And now to the remaining paper of Einstein—on quantum theory. Unlike Planck, Einstein realized at once the boldness of Planck's work. So he titled his paper "On a Heuristic Viewpoint Concerning the Production and Transformation of Light." "Heuristic" means, conducive to discovery, possibly untrue but stimulating. The production and transformation in Einstein's title is, of course, reference to Kirchhoff's law. "Heuristic" refers to the idea of light particles. The most characteristic quality of Einstein, I think, is not his insight—Newton was better at that—nor his utter simplicity—Archimedes takes the lead here—but his light foot. He was always tentative, always quick to seek contradictions between two theories, and, as far as he was in charge, to plant the contradictions between his theories from the start, so as to show that the older theory is a mere special case of, and a mere approximation to, the newer theory, thus opening the road to a crucial experiment between them.

In 1905 he established inconsistencies between
- (a) the constancy of the speed of light and Newton's theory
- (b) thermodynamics and statistical mechanics,
- (c) Planck's theory and Maxwell's theory.

This last point is where we are at just now.

I hope you remember that from the start Planck declared both Maxwell's theory and thermodynamics universal and true. But this made little impression on Einstein. He knew where to admire Planck and where to take him lightly. Allow me to tell you again that they soon became friends for life even though the older thinker was conservative, demanding to use quantum theory as sparingly as possible and defending German nationalism, while the younger thinker was radical, recommending the reckless application of quantum theory and of statistics everywhere and advocating socialism and world government. After the war Einstein was appalled by the dishonesty which his German colleagues exhibited but he never had a harsh word for Planck. Einstein considered open-mindedness a rare quality and he found it in Planck. Already in 1911 Planck stated that he found Einstein's readiness to quantize light everywhere as a radical move not to his taste, but he said this in the friendliest manner possible, and without the resentment to opponents that can be found in his scientific autobiography. Back to our story.

Einstein showed errors and incompleteness in Planck's work. He showed that Planck's equation leads not to Planck's distribution but to Wien's, and only in 1924 Bose and he replaced Boltzmann's distribution with a new, quantum one, that yielded not Wien's distribution but Planck's—thus finally closing the issue of the radiation in a black cavity, in line with his 1905 proof that Maxwell's equations plus Boltzmann's theory contradict Planck's formula. He also insisted already in 1905 on the correctness of Rayleigh's distribution (so I should not have called it a red herring or silly after all!) while preventing its catastrophic result nonetheless: if Maxwell's electromagnetic theory is given up for very weak radiation, and if weak radiation is quantized instead, then energy cannot go on leaking to each degree of freedom since the leak must be of a minimum size above the level of energy available.

This idea became central in the whole of the history of quantum theory, beginning with Bohr's theory and continuing to this day: Transitions follow quantum rules and so, even if the classical theories require that they occur they are often suppressed. These rules are called quantum exclusion rules or quantum prohibitions, and the general idea of their role and significance is (rightly) ascribed to Bohr, though already in 1905 Einstein used it. In 1905 this was not clear at all; then the chief impression received was the fierce havoc that Einstein had managed to create. From this grand demolition job he rescued the idea of light quanta possessing energy according to Planck's formula, $E = h \times \nu$, and showed that, when light is in small concentration, the deviation from Maxwell's theory is considerable, but when light is concentrated, it shows increasing agreement with that theory.

This is all; and it is terrific (even though lots of valid objections to it did develop later on), and should have excited everyone's admiration. Einstein added to this a long list of known facts explained by the quantum hypothesis. The photoelectric effect stands out: if light is distributed evenly, then a very weak beam should either not enable electrons to escape from metals, or it should take a long stretch of time—the weaker the longer—and then release a torrent of them. In fact the emission is almost instantaneous and, as Lenard had shown, each electron that escapes travels with about the same velocity if the incident light is monochromatic. This was Einstein's prediction that won him a Nobel prize: since light comes in pellets of energy, the kinetic energy of every electron emitted from a piece of metal due to the absorption of monochromatic light is that energy minus the energy it takes to tear the electron from the metal—the electron's binding energy. The reason he was awarded his prize for that is that the Swedish Academy (which is in charge of distributing the prize) was (and still is) saturated with scientists profoundly convinced of the truth of (Baconian) empiricism which rightly leads them to the conclusion that revolutions in science are impossible, and that it is safer to honor people for their solid, factual discoveries than for their questionable ideas.

The ratio between the emission-coefficient and the absorption-coefficient was soon pushed aside, the black body too became unimportant, and the quantum of light started going places on its own. In particular, all phenomena showing any threshold effect whatsoever began to make sense and could be calculated. But also remote phenomena, such as that of specific heat of solids, were better explained by Einstein.

Not that Einstein had success here. Somehow, unlike Planck, Einstein went on cheerfully and worked hard regularly. In his old age Einstein said that between 1905 and 19113 he tried hard but fruitlessly to work on emission and absorption. It was because of this, he said many years later, that he could all the more appreciate the success of young Niels Bohr's theory of the harmony of the atoms of 1913.

CHAPTER 6

6.7 The Crisis in Physics

This is the end of my story. Regrettably, many people direct their sense of nostalgia to a period just prior to their birth. For me it is the period between the rise of quantum theory and its victory. So I must stop before I make a fool of myself and gloat like ever so many historians over the greatness of the past. But, I hope you have noticed, I hold a special fondness for the period preceding that of the great explosion of modern physics, the time of crisis.

The change was great. Things started moving fast. Whittaker wrote then a report to the British Association in which he said that any page on which the printer's ink is dried is already out of date. This was in 1913. The spirit of Einstein and the exuberance of Bohr spread like a flame, a wild flame, but the crisis in science was an agony. It was no small matter to scatter to the wind the cherished ideas of the solidity and the finality of science. It is not that I like the idea of solidity. I particularly dislike the view of science as a substitute religion, a substitute more reliable and thus more comforting than the original, and also more rational because it is demonstrated. Yet I am not insensitive to the fact that this idea was very hard to give up, that after the crisis was over many a scientist was converted to irrationalism plain and simple. And Newton's theory and its like were so many fixtures of the intellectual surrounding that it was quite bewildering to see them go. So much that had been taken for granted was shaking and most people had no training to help them under such conditions.

Atoms split. Small issues began to grow and upset science, the absent ether wind and the enigmatic black cavity being the most prominent among them. New instruments came in and looked exciting but also frightening. You must have seen old movies on the late show, science fiction and science horrors, with laboratories full of electric machines and glow tubes and lights and chemicals bubbling. These laboratories are fairly authentic late nineteenth-century imageries built by artists who had seen them with their own eyes. They meant to capture—and exaggerate, of course—a new mood, a new mistrust of science, which, alas, proved only too justified. Not only were x-ray machines at once so useful; during the first years of application they burned people and for many decades their application caused cancer. Once discovered, the defects were eliminated, of course, but this is not always easy. And then came nuclear explosion. This is not an accusation but an attempt to revive for a brief moment the bewilderment science has caused: science had promised too much comfort, too much stability, too much proof, rationality, and benefit to all. The crisis of science was a case of bankruptcy. Poincaré and Duhem said, scientific change could be controlled. Most scientists recommended caution.

And then came the revolution. Nobody got hurt from the revolution; least of all science itself. But two alternative ideas were beheaded: that science endures and suffers no change and that all scientific change is small. Increasingly the need to readjust to scientific change is realized. Clearly, great changes in science are exciting but are hard to produce. Kuhn's central critique of Popper's philosophy of science is just this: revolutions, he says, are very nice but life is not all pie. This is in line with the idea that a revolution or two in a century will do, an idea which is still advocated by some popular stuffed-shirt professors of science. This is like saying, since life is not all pie, let us make sure that people do not have much fun. This is so futile, because as life is not all pie, there is no need to insure that we will have too much fun. What the stuffed shirts mean is that we should have less fun than possible. Why? Because science needs some stability, they say, and too many revolutions bring about chaos. Perhaps; let us then have as much fun as possible, and when the result begins to be too chaotic for our own good, we will do something to curb our appetite for fun. After all it is the fun of scientific progress. As for me, I still hope to see another explosion of science; it is not so much an internal affair—who knows what are science's internal affairs and these anyway can go zigzag—but a matter of willingness to be daring. Hopefully; I cannot foretell. All I have tried to show is how the story developed to the point of explosion; the quantum explosion itself will still engage many writers and many readers for years to come. It may still inspire some young bright-eyed dreamers to dream of the next explosion, of the next scientific revolution.

Appendix A

The Kirchhoff-Planck Radiation Law

It is well known that Planck studied Kirchhoff's radiation law because he was attracted by its utter generality, and that he was thus led to his theory of the quantization of light. Why this stress on utter generality? Did Wien, for instance, in studying Kirchhoff's law, disregard its generality? Are not all laws of nature general, and attractive on account of their generality? [1]

Moreover, how general, precisely, is Kirchhoff's law? One may say it applies to all thermal radiation of all black bodies which are in equilibrium with their environments. The previous sentence contains three restrictions: the radiation has to be thermal, the radiating body black, and the setup that of thermal equilibrium. We may omit any one of these restrictions and obtain three different interpretations of the law; we may omit any two of these restrictions and obtain three more interpretations; and we may omit them all. Thus we can have at least eight (empirically) different interpretations of the law. I say at least eight different interpretations because we may also interpret terms like "thermal radiation" differently. The most radically narrow interpretation of this term will be (intentionally) circular: that radiation is thermal which obeys Kirchhoff's law. This interpretation is so narrow that, once we restrict the law to thermal radiation in this interpretation of the word, we obviously do not have to restrict the law any further; it becomes an immediate corollary to the definition and thus trivially a tautology. This is, indeed, how *Handbuch der Physik* introduces the law [2, p. 133]. There is, however, an elaborate proof of the law [3], in which the second law of thermodynamics is used, and so, evidently those who accept the proof accept a different interpretation of the law. One might expect—quite *a priori*, indeed—to be able to read from the proof the right interpretation. In fact, however, all eight interpretations mentioned above are found in the introductory literature, even in the leading introductory literature [4], not to mention works which are ambiguous about this point, or even inconsistent. I do not speak of marginal works; I need not discuss the importance generally and quite rightly attached to Georg Joos's book [5], yet it is plainly inconsistent on the point of interpretation, or at least it is sufficiently ambiguous to be so read.

The reason for this diversity of interpretations and confusions may lies in the fact that the proof of Kirchhoff's law is not clear. Inasmuch as this is the case, I hope my present restatement of it will dispel some of the obscurity. I am afraid, however, that there is a little more to it than this. Young students are often bewildered not only by questions concerning the generality of Kirchhoff's law but also by the strange fact that the law is mentioned very seldom in the literature—that even Kirchhoff's emission and absorption coefficients are seldom mentioned—despite the fact that the law was proposed at so crucial a juncture as the time of the rise of quantum theory. Browsing through the literature, one may find an occasional use of Kirchhoff's law in some experimental physics, but the only place where it is treated at all seriously today is in the astrophysical literature. Thus, Chandrasakhar presents the Kirchhoff-Planck law very clearly though as a mere approximation to better laws [6]. Here, indeed, is the crux of the difficulty. As I have said elsewhere [7, 8], often writers cannot repeat an idea which has been superseded without some tinkering; the tinkering does not improve matters, but it does often obscure the reasoning which underlay the old idea when it was proposed. (*Handbuch der Physik* at least claims that Kirchhoff's own presentation was unsatisfactory.) Historians, however, cannot allow themselves the luxury of improving upon history. Let me, then, present my view of the history of the case as briefly and schematically as I can.

A.1 Prevost's Law of Exchange

That radiation may be related to heat is ancient knowledge. Blacksmiths from time immemorial knew how to compare the temperature of metals by the color of their radiation. Yet it was Newton who declared all radiation to be a function of the temperature of the radiating body; he declared that fire is but radiating hot gas. [9] To illustrate this he filled a glass tube with black smoke and heated it until the smoke glowed. [9] Newton's illustration was not conclusive, of course. In particular, the following objection may easily be imagined. It might be claimed that the smoke radiated in his experiment not from being heated by its environment but from some other environmental influence, for example, from being forced to absorb fire atoms. The claim that Newton was mistaken may be supported by the observation that, once the heated glass tube is removed from the environment, it ceases to radiate after a very short period. It may be claimed that, though the tube is hardly cooled, it ceases to radiate because its environment has changed.

This objection to Newton's idea was answered by Prevost, who postulated, in 1809, that emission is cooling and absorption is heating. [10] It follows from this postulate that, under obvious circumstances, one body may emit radiation and thereby cool and cease radiating, and that the radiation it has emitted may be absorbed by another body, which may thereby heat and start radiating light, which will be reabsorbed by the first body, and so on periodically. This idea is known as Prevost's law of exchange. [11] Thus far I have presented a crude idea concerning the relations between emission and absorption, one only incidentally related to the problem of the nature of fire. The problem of the relation between absorption and emission became significant with the discovery of absorption spectra and the rise of astrophysics—or, more precisely, solar physics (and solar chemistry).

A.2   Fraunhofer's Discovery of Spectroscopy

Spectral lines were observed a few times in the 18th century, but no significance was attached to these observations and they were not generally known. The real discovery of spectral lines is therefore attributed to Wollaston (1802), who thought they were boundaries between colors and who devised a method of observing them [12], and to Fraunhofer (1817), who is the father of astrospectroscopy. [13] The story of its development goes this way.

Newton had thought that the spectrum contained seven colors, with ranges having arithmetic ratios corresponding to a Pythagorean scheme of harmony. He confessed his own inability to see borderlines between color ranges but claimed that an assistant of his could see them rather systematically. [14] Next, Thomas Young started his studies by examining contemporary acoustic theories and trying to relate harmony to physiological acoustics. [15] He turned to physiological optics as an after-thought and failed to develop a satisfactory physiological optics of seven colors. Most of the historical literature today criticizes Newton's corpuscular theory of light as defective, on the ground that contains the postulate of an infinite variety of light particles, corresponding the continuum of the wavelengths of the visible spectrum. This criticism is a hindsight. The proper presentation is this. Newton had postulated only seven kinds of light corpuscles, each having a spread over that part of the spectrum allotted to it. It was only when Young threw doubt on the existence of the seven colors that the choice was between (i) infinitely many (rather than seven) kinds of particles and (ii) one kind of wave with infinitely many different wavelengths. This is why the doubts concerning the seven colors became so crucial, and why Young turned to Huygens. [16]

Wollaston devised the first spectroscope, and with it discovered (1802) some of the solar dark lines. He thought these might be the boundaries between colors. [17] This idea was soon destroyed by Fraunhofer's discovery (1817) of the multitude of these lines. [13] Wollaston also discovered the emission spectra, and the fact that, for sodium, the absorption spectrum may be the negative of the emission spectrum. On the hypothesis that all spectral lines of the sun are negatives of emission spectra of flames on earth, chemical analysis of the sun's composition could begin.

But the hypothesis is false. The history textbooks tell us of elements, such as helium, which were first identified in a solar spectrum; they omit mention of elements allegedly discovered on the sun and later declared nonexistent. The fact is that the relation between emission and absorption spectra is not always as simple as that between a negative and a positive photograph. [18] Unless we realize this, the chief impetus for Kirchhoff's research is not noticed, and his results may seem mysterious.

A.3 Stewart's Law of Radiation

It would be very easy to complicate this study by considering all sorts of factors that destroy the symmetry between absorption and emission. Some of these factors—such as the Doppler effect—are very important for astrophysics but play almost no role, or none at all, in the present study of the relation between absorption and emission. Other factors are more significant in this context—factors such as chains of emissions and absorptions which take place when two radiating elements interact but not when each one of them is excited separately. Let us ignore all such complications and center on one factor—temperature, with which we started. One thing we have noted already: that hot bodies emit and thereby cool, and that sometimes when they are absorbing they heat up enough to radiate (this is Prevost's law of exchange).

The simplest way to understand Prevost is to say that, although a body emits only when it is hot, it absorbs whenever it is exposed to light. That this is an oversimplification is common knowledge: ordinarily bodies absorb certain wavelengths, reflect others, and transmit still others. Indeed, the idea of the reciprocity between emission and absorption attempts to take these differences into account. So let us consider just one wavelength and the material which, when hot, emits it. Now one can read Prevost's law of exchange in the following manner: matter absorbs a wavelength which, when hot, it emits. This is Ångström's law (1853). [19] As to the wavelengths which the given body cannot emit, it cannot absorb them under any conditions. This is Stewart's law (1854). [20]

One corollary to Stewart's law may be pointed out at once. Suppose there exists a body which may, under some conditions (namely, when hot), emit white light—light of every wavelength of the spectrum. Then this body must also absorb all light—it is black. A black body is one which absorbs all light which falls on it; when cold, it appears black but, when hot, a black body may radiate light of different wavelengths, including white light, and thus appear to the eye not black at all. In any case, appearance to the eye is incidental. As Foucault had shown a little earlier (1849), a small flame located between the eye and a large flame of the same kind appears dark (because its absorption effect is stronger than its emission effect, and because the eye adjusts to the setup). [21]

A black body, then, is supposedly that body which absorbs all wavelengths, regardless of whether or not it looks black. This definition is not good enough, because even transparent bodies seem to absorb some portion, however minute, of any given wavelength. And so both Ångström's law and Stewart's law look either false or hopelessly inadequate. It is easy to correct the definition of a black body. Whereas, before, we spoke of blackness as the ability to absorb radiation of any wavelength and concluded that all bodies are black, we may now speak of blackness as the ability to absorb in its entirety any radiation of any wavelength. This definition will make blackness not too common but too rare to be a subject of study. Moreover, it raises the need to distinguish between various phenomena previously lumped together. Let us consider Foucault's experiment again, with a strong and a weak radiation source of the same material, and ask whether that material may be black. If, as Foucault thought, the weak source creates an absorption spectrum by dispersing light from the strong source through resonance, then the weak source cannot be made of black matter. Black matter cannot disperse light, since it must absorb all of it. It may, however, create an absorption spectrum by absorbing light from one direction, storing it for a while, and then emitting it in all directions. This is not resonance but resonance emission.

The existence of resonance emission proper, we have seen, contradicts Prevost's law of exchange. We must, therefore, declare seeming resonance emission to be really some other phenomenon: absorption raises the temperature of the weak source for the delay period, and this slight increase in the temperature causes the increase in radiation. We thus see that both Ångström's and Stewart's laws raise more problems than they solve; they force us to speak of the degree of absorption and of the relation between absorption, emission, temperature, and so on. This was the situation when Kirchhoff entered the scene, trying to take all these complications into account. This, indeed, was his first (1859) achievement. [22]. For simplicity's sake, I present it as follows.

A.4 Preliminaries to Kirchhoff's Law of Radiation

Let us take one unit of matter of a given chemical nature; the unit may be an arbitrary volume or mass, or it may be an atom. Let us use the following, rather loose, nomenclature concerning its emission, its absorption, and its disposition to emit and to absorb during a unit of time:

$E =$ *actual emission* ; $e =$ *the disposition to emit* ; $A =$ *the actual absorption* ; $a =$ *the disposition to absorb* .

Obviously, since A depends on the environment (there is no absorption when there is no surrounding radiation, for instance) in a manner different from that in which a depends on the environment, these terms cannot be equal. We may suggest that e and a are independent of the environment, and that E, too, in accord with Prevost's law, is not explicitly dependent on the environment. We postulate the following. Functions e and a are explicit functions of internal variables only, such as the temperature of the material in question and its chemical composition. We also postulate,

$$E = e .\tag{1}$$

Equation 1 is false, since it implies that there is no induced emission. However, it is an immediate corollary of Prevost's law, and so we shall pretend that it is true. Furthermore, we postulate

$$A = I \times a ,\tag{2}$$

where I is the intensity of the incident beam or the density of the radiation in the immediate environment. One immediate corollary from Equations. 1 and 2 is of supreme importance, and takes much of the mystery out of the dimension of Planck's constant, as we see later. (Dimensions are denoted here by square brackets.) We have seen that, by definition, $[E] = [A]$; and from this we can deduce, with the aid of Equations 1 and 2, that

$$[e] / [a] = [E][I] / [A] = [I]$$

which is the dimension of the density of light passing through a unit of area in a unit of time t, or the dimension of energy divided by $L^3 \times T$.

Equation 2 is false, too, since it assumes that excited atoms have the same disposition to absorb as unexcited ones. But the merit of Equation 2 is its simplicity—that is, its linearity—and so it is a good approximation in weak fields; we shall accept it for the time being as true.

Up to now we have been very imprecise, speaking of e, a, and so on without paying heed to wavelengths at all. So let us correct this and speak from now on of e_λ, a_λ, E_λ, A_λ of the given unit of matter; we shall retain Equations. 1 and 2 with this modification. Equation 1, now $E_\lambda = e_\lambda$, has changed its meaning in a somewhat unintended way: it now denies not only the possibility of induced emission but also the possibility of resonance emission. Similarly, Equation 2, which becomes $A_\lambda = I_\lambda \times a_\lambda$, now denies not only saturation but also secondary absorption. Still, we shall accept both equations as true. Now, Stewart's law becomes:

$$e_\lambda \equiv 0 \;\rightarrow\; a_\lambda \equiv 0. \tag{3}$$

In other words, what a body cannot emit (when hot) it cannot absorb (when hot or cold); we may rewrite Equation (3) thus:

$$E_\lambda = K_\lambda \times a_\lambda \quad \text{where} \quad K_\lambda \geq 0 \tag{3a}$$

for all wavelengths, where K_λ is a still undefined coefficient, except in that it is nonnegative. Both Equations (3) and (3a) contradict our observation that every body will absorb some portion of any incident radiation, that no body is utterly unable to absorb any given wavelength. From the viewpoint of the old quantum theory it is quite clear that some quanta will not be absorbed by some systems under any conditions; but in the old quantum theory the term *"absorption"* is much more rigorously defined than the term *"absorb"* is in the observation that any matter will absorb some portion, however small, of any incident radiation. The situation here is much more baffling than it looks, and we must, in desperation, declare Equation (3a) to be true and turn a blind eye to many common observations as being too complex to handle as yet. So, back to Equation (3a).

Since e_λ and a_λ are both explicit functions of internal variables only, so is K_λ. For, by the very use of Equation (3a), K may be construed as an explicit function of internal variables. To put it differently if K were an explicit function of an environmental variable, then, obviously we might, by varying that variable, alter e/a.

To take care of Ångström's law, we must view a_λ as larger than or equal to e_λ at low temperatures and the reverse at high temperatures, or else view both e_λ and a_λ as 0. In the latter case, it does not matter what value we assign K_λ. Otherwise, we can write

$$K = e/a. \tag{3b}$$

remembering that the arguments of the equation are all internal variables.

The standard way of writing Equation (3b) is

$$e(\lambda, T, \ldots) / a(\lambda, T, \ldots) = K(\lambda, T, \ldots) \tag{3c}$$

where the dots stand for the unspecified internal variables, such as chemical situation, specific gravity, or specific heat. The change from e to e_λ was viewed as an increase in precision; the change from $e_\lambda(T \ldots)$ to $e(\lambda, T \ldots)$ is viewed as mere change of nomenclature. But Equation (3c) may somehow have smuggled in the hypothesis that K for any single wavelength is independent of K for any other wavelength—that a body's absorption-emission mechanisms for one wavelength (the energy levels for one wavelength) bear no relation to those for another wavelength. The Bohr model already assures us that this hypothesis simply is not true. But we have not advanced thus far yet. Also, we may say that if there really is any interdependence between mechanisms for absorption and emission of different wavelengths, then the dots in Equation (3c) will register that difference. Moreover, the dependence of the quantities in Equation (3c) on temperature may take care of the point at issue: remembering Prevost's law we may say that the body in question, when absorbing wavelength 1, is heated, and so it may thus change its pattern with respect to wavelength 2. Are there other internal variables significant in this way, and, if so, what are they? If not, how does the dependence on temperature reflect the absorption-emission mechanism?

A.5 Kirchhoff's Law and Its Proof

Kirchhoff's most important step was to prove the following formula:

$$K(\lambda, T, \ldots) = K(\lambda, T) \tag{4}$$

where, again, the dots stand for unspecified internal variables.

The proof can be given in two steps. First, imagine that,

$$\textit{under some external conditions}, \ K(\lambda, T, \ldots) = K(\lambda, T) \tag{5}$$

APPENDIX

Since K is independent of external variables, Equation (4) follows from Equation (5). (Equation (4) may read, "*under all external conditions, etc.*" whereas Equation (5) reads, "*under some external conditions, etc.*" Yet the stronger Equation (4) follows here from the weaker Equation (5)!)

The second step is to prove Equation (5). Take two bodies amalgamated with a cavity between them, both having temperature T and both emitting a wavelength λ. If K is not the same for both under such equilibrium conditions, radiation may be used to destroy the equilibrium, contrary to the second law of thermodynamics. For, this law implies that a system does not by itself move away from thermal equilibrium. Hence under equilibrium conditions, K is the same for both bodies, and so Equation (5) is true and thus our proof is complete.

There are various more precise versions [3, 22] of the proof of Equation (5); what is objectionable in the proof, however, is not its imprecision but the assumption it rests on, which is much more stringent than appears to be the case. The assumption is that any two bodies at any temperature may radiate any given wavelength, which we may filter and allow to be emitted and absorbed by both bodies in isolation from any other wavelength, interference from the filter (in terms of temperature and of emission of its own radiation), and so on.

Let us accept the proof, nonetheless—namely, accept

$$e/a = K(\lambda, T)$$

as a universally true equation. Notice that we have no reason to accept $e = e(\lambda, T)$ or $a = a(\lambda, T)$; indeed, were e independent of other variables, spectroscopy would be impossible. To estimate K we may choose a black body. By definition, a black body is a body for which $a \equiv 1$; thus,

for a black body, $e = e(\lambda, T) = K(\lambda, T)$.

Are there any black bodies? Can we assume that any exist? If the assumption that they do leads to a theoretical difficulty, the difficulty may be used as proof for the nonexistence of black bodies. [23] We shall, therefore, have to avoid all theoretically significant corollaries to the hypothesis that black bodies exist; the use of black bodies is either a simplification of a discussion for the sake of convenience or an asymptotic ideal case considered for the sake of empirical investigation. Let us assume the existence of a black body and the presence of a cavity in it, under equilibrium conditions. We can easily see that, for any given wavelength, there is in the cavity a constant energy density ε of waves of that length, and that, in each time unit, each part of the walls of the cavity absorbs this amount of energy multiplied by c, the speed of light, and emits the same amount of energy. That is,

for a black cavity, $E = A = I = \varepsilon \times c = e/a = e = K$.

If we fix T, make in the black body a very small hole leading to the cavity, and place in the hole a filter for light of fixed λ, we can measure I, provided our interference with the system is negligible. We may consider an alternative setup. Einstein has envisaged a perfectly white cavity ($a \equiv 0$) with a piece of a black body inserted for a while to arrange the energy distribution in the cavity to equal K. For this white cavity (with the black body eliminated), K is of no interest, since, if a equals 0, e also equals 0 and K has no physical significance; yet this white cavity will give us a way to measure K and then use the results on ordinary bodies where K is of great significance, especially where we are trying to deduce emission spectra from absorption spectra.

A.6 Between Kirchhoff and Planck

The Stefan-Boltzmann [24] law (1879) is as follows:

$$I(T) = \int_0^\infty K(\lambda, T) \, d\lambda = \sigma T^4. \tag{6}$$

Boltzmann derived it [25] from the Maxwellian assumption that radiation causes pressure, and from the two laws of thermodynamics. It is nowadays considered valid for black bodies only, since, now, K is viewed as being less universal than Kirchhoff thought it was; but Boltzmann's considerations are not hereby impaired. The same thing may be said of Wien' theorem (1893) [26, 27],

$$K(\lambda, T) = \lambda^{-5} f(\lambda, T) \tag{7}$$

where f is a still undetermined function of the one variable $\lambda \cdot T$

$$f(\lambda, T) = f(\lambda \cdot T).$$

Equation (6) follows from Equation (7), as it should.

Also, by considering a parallel between the radiation pressure and the pressure of a Maxwellian gas, Wien further determined the relationship known as Wien's law (1896) [27]:

$$f(\lambda, T) = e^{-a/\lambda T} \tag{8}$$

This parallel was made complete in Einstein's famous paper of 1905, where radiation was itself treated as a gas proper (quantized).

From considerations of degrees of freedom and the Boltzmann distribution, Lord Rayleigh concluded (1900) [28, 29], however, that

$$f(\lambda, T) = A \lambda T. \tag{9}$$

That Equation (9) violates the idea of equilibrium between radiation and the wails of the cavity is obvious; this follows from the assumption of a Boltzmann distribution for infinitely many degrees f freedom, as Jeans pointed out in 1901 and. 1904 [30], on quite general grounds. He even quoted Maxwell to say that the introduction of degrees of freedom for radiation brings in infinitely many new degrees of freedom, and that hence any distribution which takes them into account will lead to a catastrophe. Jeans himself thought that, if the time required for its occurrence were very long, the catastrophe would not matter overmuch. [31]

Planck was quite ignorant of this difficulty and even of the very existence of Rayleigh's formula [32, 33]—for reasons mentioned below—whereas Rayleigh was well aware of Planck's formula and, in a note published in 1902, refers to it and to the simple relation between the Planck and the Rayleigh formulas. [29] That he did not follow Planck's reasoning, then or later (as he confessed in 1911 in a letter to Ehrenfest) [33] is a different matter. Planck's reasoning still requires some clarification.

A.7 Planck's Studies Prior to His Quantization

Enter Planck.

Planck's initial interest was in thermodynamics. He interpreted [34] Helmholtz's view on the law of conservation of energy in a very interesting manner, as follows. Given the forces acting in a system, we can examine them, or the potentials from which they may derive, and deduce the law of conservation of energy. We may assume, quite generally, even if we do not know the forces in a system, that these forces are conservative. It may be argued that such an approach is defective, since it is based on ignorance; not so, says Helmholtz according to Planck; the law of conservation of energy is philosophically deeper than any law of force of any system; conservation of energy is the primary law of nature. Planck greatly admired Helmholtz, and he intended to apply to the second law of thermodynamics the same mode of reasoning he thought Helmholtz had applied to the first law. Whereas others would view the application of these laws as a kind of shortcut to circumvent our ignorance, Plank viewed these Laws as primary. Whereas others preferred, when possible, to deduce the two laws from specific conditions of given systems—whether laws of force, models, distributions, or other conditions—Planck preferred, to deduce his results from the laws without specific assumptions (which he called "models" or "mechanisms", in this generalized sense) whenever possible. Whereas Lorentz derived the theory of oscillators from this theory of the electron, Planck tried to deduce it from Maxwell's equations and from general considerations only. He said later that he was unfamiliar with Lorentz's work, but that he would have rejected it anyway as being based on a

APPENDIX A

specific mechanism [35]. And he refused to assume the Maxwell-Boltzmann distribution throughout his work; he did assume it in the last stage of his work, but he never viewed this as anything but a symptom of a defeat. [36] Planck's interest in Kirchhoff's law derived from its utter universality, which suggested that it depended on no specific mechanism. [37] He was dissatisfied with Wien's derivation of the value of $f(\lambda \cdot T)$ which, we remember, was based on an analogy with a Boltzmann gas—a dual violation of Planck's fundamental principles (since it was an analogy, and an analogy to a model). Planck's first contribution was to correlate the average energy of an electrodynamic (Maxwellian) oscillator and the average energy of the field with which it is in equilibrium, for a given frequency, or, rather, for the frequency range $(\nu, \nu + d\nu)$. His formula, which Lorentz had arrived at by other means, is

$$E_\nu = 8\pi\nu^3 / c^3 \cdot U_\nu \, d\nu \tag{10}$$

where U is the average energy of the oscillator for the given frequency range and E is the energy density of the field for the same range. (Note that $[E_\lambda] = [U_\nu]/L^3$.)

Planck's second step was in line with exercises which he performed, and described in his *Treatise on Thermodynamics*, for various systems: he searched for an arbitrary function S such that the second law of thermodynamics would be characterized by its asymmetry and such that S would help describe the behavior of the system under consideration—in the context of this discussion, would help in the derivation of Wien's law. First we have to translate Wien's theorem and Wien's law from a function of K to a function of ν, remembering that

$$d\nu = d\left(\frac{c}{\lambda}\right) = \left(-\frac{c}{\lambda^2}\right) d\lambda$$

or

$$d\lambda = -\frac{\lambda^2}{c} d\nu .$$

Equation (7) then becomes

$$K(\nu, T) = \nu^3 f(\nu/T) \tag{7a}$$

where we write $f(\nu/T)$ in preference to $f(T/\nu)$, because this makes it easier to write

$$f(\nu, T) = e^{-a\nu/T} \tag{8a}$$

By Wien's law, the energy density in a black cavity is

$$U_\nu = \alpha \nu^3 e^{-\beta \nu / T} \tag{11}$$

and, from Equation 10, we have

$$U_\nu = \gamma \nu e^{-\beta \nu / T}. \tag{12}$$

From thermodynamics we have

$$dS = dU / T$$

or

$$\frac{dS}{dU} = \frac{1}{T},$$

and from Equation 12 we have

$$\frac{1}{T} = \frac{1}{\beta} \ln \frac{U}{\gamma \nu} \tag{12a}$$

so that

$$dS = \frac{dU}{T} = \frac{dU}{\beta \nu} \ln \frac{U}{\gamma \nu}$$

and

$$\frac{d^2 S}{dU^2} = \frac{\delta}{U}. \tag{13}$$

Under the force of experience, which refuted Wien's law [see 38], Planck changed his Equation (13) to

$$\frac{d^2 S}{dU^2} = \frac{a}{U(b+U)}, \tag{14}$$

from which he concluded, with the aid of Wien's theorem and his own law for the relation between the energies of the oscillators and the cavity (Equation 10) that

$$U_\nu = h\nu \, (e^{-h\nu / kT} - 1)^{-1} \tag{15}$$

and

$$E_\nu = \frac{8\pi h\nu^3}{c^3} (e^{-h\nu/kT} - 1)^{-1} d\nu. \qquad (16)$$

It is a strange fact that Planck's discovery of the constants h and k was made prior to his quantization of radiation, and that he himself and tradition have obscured this fact. [39] Planck was so impressed by these constants and their dimensions that, even before he introduced quantization, he spoke of the new possibilities that he had opened toward a theory of natural constants and natural units.

A.8   Einstein's Version of Kirchhoff's Law

Planck's work on the entropy of oscillators has been, of course, entirely superseded by the work of others. Its historical importance is that it enabled him to modify Wien's formula (Equation 9) by modifying Equation 13 into Equation 14 by the use of Equation 10. The strange dimension of h in Equation 16 looks strange only in retrospect; there was nothing puzzling about it before Planck's postulation of his famous quantization of energy.

Planck set out to solve one problem and ended up with three. His original problem of explaining Kirchhoff's law was one; the nature of quanta and the explanation of the distribution of radiation in the cavity are the two new ones. All too often people confuse Kirchhoff's law with the law of distribution of radiation in black cavities under equilibrium conditions, because of the numerical equivalence of these two laws. In any case, it is well known that Planck thought he had explained the black-cavity radiation distribution by quantization plus Boltzmann distribution. These postulations lead to Wien's law, not to Planck's ; [40-42] replacing Boltzmann's distribution by the Bose-Einstein distribution will lead to Planck's law of distribution. But Kirchhoff's law of emission and absorption is still unexplained; though Planck's distribution does provide the estimate for the correlation between emission and absorption, it does not explain it. The next development of quantum theory was divorced from all thermal considerations and all statistical considerations, but it did throw immense light both on quantization and on the relations between emission and absorption; I am referring to Bohr's theory, of course, which needs no discussion here. But the step following it deserves some mention, since it is all too often neglected, and since, when represented, its logic is not made clear. [43] I refer to Einstein's theory of Kirchhoff's law of 1916. [41, 44]

Whether because of the studies of induced emission (resonance radiation, fluorescence) of Wood and of others [45] or *ad hoc*, Einstein, in order to arrive at Planck's distribution, assumed the existence of induced emission. Further, he considered the dispositions of matter to emit and absorb definite quanta to be in accord with Bohr's theory. Thus, instead of speaking of E_ν, he spoke of $E_{m,n}$, the disposition of matter to emit by a quantum jump from energy level m to energy level n, and, similarly, replaced A_v, with $E_{m,n}$; but he did not assume $E_{m,n}$ to equal $e_{m,n}$, as $e_{m,n}$ depends on internal variables alone, whereas $E_{m,n}$ depends on the field, too, since induced emission is allowed.

Again, let us consider a black cavity under equilibrium conditions [see 46]. Due to the equilibrium conditions, the probability of an oscillator's attaining the energy state U_m, is

$$p(u_m) = e^{-\varepsilon_m/kT}. \tag{17}$$

The probability of emission $v_{m,n}$ when the state m occurs is denoted as $A_{m,n}$; the probability of emission $e_{m,n}$, then, is

$$e_{m,n} = e^{-\varepsilon_m/kT} A_{m,n}. \tag{18}$$

The absorption coefficient $a_{m,n}$ is likewise defined, so that

$$a_{m,n} = e^{-\varepsilon_m/kT} B_{m,n}. \tag{19}$$

where $B_{m,n}$ denotes the probability of absorption of $v_{m,n}$ when the state n occurs and $v_{m,n}$ is present. In line with the linear Equation 2 above, let us assume that

$$A_{m,n} = a_{m,n} E(v_{m,n}, T) \tag{20}$$

where E is the field-energy distribution, as above.

The third factor to consider now is induced emission, where, instead of $E = e$, as postulated above, we postulate $E = e+i$, where i denotes induced emission and

$$I_{m,n} = e^{-\varepsilon_m/kT} C_{m,n} \cdot E(v_{m,n}, T), \tag{21}$$

where $C_{m,n}$ denotes the probability for induced emission of $v_{m,n}$ when the state m occurs and when $v_{m,n}$ is present. Since, in equilibrium, $A_{m,n} \equiv E_{m,n}$, clearly,

$$e^{-\varepsilon_n/kT} B_{m,n} \cdot E(v_{m,n}, T) = e^{-\varepsilon_m/kT} A_{m,n} + e^{-\varepsilon_m/kT} C_{m,n} \cdot E(v_{m,n}, T) \tag{22}$$

It is nice to note that the linearity in the various equations, though used more extensively than in the original considerations, is now very natural, since it follows from assumptions of stochastic independence which are quite natural. We now introduce a new assumption, which is physically appealing, though it may be less neat. It is the assumption that when T goes to infinity, E does too. Eddington has shown that this assumption is unnecessarily strong—that monotony will do. [47] Nonetheless we

APPENDIX A

shall make the strong assumption, which reduces Equation 22 with T going to infinity to the case

$$B_{m,n} = C_{m,n}. \tag{23}$$

From Equations 22 and 23 we easily deduce that, under equilibrium conditions,

$$E(\nu_{m,n}, T) = (A_{m,n}/B_{m,n}) \cdot \left[e^{-(\varepsilon_n - \varepsilon_m)/kT} - 1\right]^{-1}. \tag{24}$$

Postulating, with Bohr, $\varepsilon_m - \varepsilon_n = h\nu_{m,n}$, and, further,

$$A_{m,n} = [8\pi h \nu^3 / c^3] B_{m,n} \tag{25}$$

we can identify Equation 24 with Planck's distribution (Equation 16). It should be noted that the only distribution assumed concerned the oscillators, not the field, and that the general assumption was that of equilibrium conditions. But Equation 25 as well as the corollary, Equation 23, may be generalized, because they contain no explicit reference to the equilibrium conditions. In this respect Einstein was using Kirchhoff's techniques; but we have thus far not achieved anything like Kirchhoff's law.

It is true that Equation 25 entails a modern version of Stewart's law, but we do not need Einstein's theory for that; Einstein takes for granted Bohr's theory of energy levels, from which that version of Stewart's law follows. The task now is to word Kirchhoff's law with the aid of Einstein's coefficients. In the first place, Einstein's formula for Planck's distribution evidently does not necessarily hold for non-black bodies or for non-equilibrium systems. Moreover, the ratio $A_{m,n}/B_{m,n}$ which equals the ratio $e_{m,n}/a_{m,n}$ is very far from being e_λ / a_λ, and, as $E_{m,n}'/A_{m,n}$ tends, under equilibrium conditions, with strong fields, to approach $A_{m,n}/B_{m,n}$, it is a poor substitute for e_λ / a_λ.

To be precise, we should sum over all m and n so that $E_m - E_n$ has the same value $h\nu$ for any given ν. Thus,

$$K = \frac{\sum E_{m,n}}{\sum A_{m,n}} = \frac{\sum p(U_m)[A_{m,n} + B_{m,n} p(\nu)]}{\sum p(U_n) B_{m,n} p(\nu)}, \tag{26}$$

where $p(U_m)$ is the probability of an oscillator's being in energy state ε_m, and is an explicit function of T, and $p(\nu)$ is the probability that a quantum of frequency ν is sufficiently near to the oscillator to interact with it, and is a function of the field distribution and intensity.

We do not yet know how to assess these probabilities under any conditions other than those of a black cavity and equilibrium. At all events,

$$K = \sum \frac{p(U_m) A_{m,n} [1+\alpha p(\nu)]}{p(U_n) \alpha A_{n,m} p(\nu)} \tag{26a}$$

where

$$1/\alpha = 8\pi h\nu/c^3$$

as in Equation 25, or

$$\sum \frac{p(U_m)}{p(U_n)} \left[\frac{1}{\alpha p(\nu)} + 1 \right]; \tag{26b}$$

and, assuming equilibrium conditions and applying Equation 17,

$$K = e^{-h\nu/kT} \left[\frac{1}{\alpha p(\nu)} + 1 \right]. \tag{26c}$$

And so, again, Einstein's coefficients are eliminated from the Kirchhoff formula, as they were from the Planck formula. So long as $p(\nu)$ is very large, we may view K as dependent solely on the probability that the oscillators attain energy states U_m and U_n; these probabilities may be functions of the temperature alone, in collisions, or of $p(\nu)$ in the case of resonance radiation. Still, it is quite obvious that here K is not in the least universal even when $p(\nu)$ does not enter the picture at all.

What has become of the proof that, in equilibrium, K must be universal or else the second law of thermodynamics is violated? This proof is still valid, but only as an approximation for equilibrium conditions. According to Einstein's theory, less of the equilibrium theory can be generalized than could be generalized under Kirchhoff's, since the division into internal and external variables—indeed, Prevost's law of exchange—has long since been dropped.

A.9 References and Notes

1. For generality in science, see K. R. Popper, *The Logic of Scientific Discovery* (1959), sect. 18. For Planck's stress on the generality of Kirchhoff's law, see his Nobel Prize lecture, *The Origins and Development of Quantum Theory* (1922), p. 3, and his *Scientific Autobiography* (1949), p. 34. The stress is repeated by Klein [M. J. Klein, *Arch. Hist. Exact Sci.*, 1, 460 (1962); 9 in *The Natural Philosopher*, D. E. Gershenson and D. Greenberg, eds. (1963),I, 96]. It is also often repeated in the

APPENDIX A   133

introductory physics literature [see, for example, F.K. Richtmyer and E.H. Kennard, *Introduction to Modern Physics* (ed. 5, 1956), p. 1471.

2. S. Flügge, ed., *Handbuch der Physik*, vol. 26 (Springer, Berlin, 1958).

3. J.H. Poynting and J.J. Thomson, *A Text-book of Physics*, vol. 3: Heat (London, 1904), Chap. 15; E. Pringsheim, *Verhandl. Deut. Phy. Ges.* 3, 81 (1901); L. Dunoyer, *Ann. Chim. Phys.* 8, 30 (1906); G.C. Evans, *Proc. Amer. Acad.* 46, 97 (1910); D. Hilbert, *Z. Physik* 13, 1057 (1912); discussion between Pringsheim and Hilbert, *ibid.* 14, 589 (1913); *ibid.*, p. 592; *ibid.*, p. 847.

4. For the most restricted interpretation, see M. Born, *Optik* (Berlin, 1933), p. 460, and F. K. Richtmyer and E. H. Kennard, *Introduction to Modern Physics, op. cit.* For the unrestricted interpretation, see S. Flügge, ed., *Handbuch der Physik*, vol. 20 (Berlin, 1928), p. 122 ff., and E.A. Jenkins and H.E. White, *Fundamentals of Optics* (ed. 3, 1957), p. 430. The equilibrium condition alone is commonest, its *locus classicus* for general physics being M. Planck, *Introduction to Theoretical Physics, Theory of Heat* (London, 1932), p. 185, as well as Planck's *Vorlesungen über die Theorie der Wärmstrahlung* (Leipzig, ed. 5, 1923); or, for astrophysics, S. Chandrasakhar, *An Introduction to the Study of Stellar Structures* (Chicago, 1939), p. 203. See also Einstein's scientific autobiography in P.A. Schilpp, ed., *Albert Einstein: Philosopher-Scientist* (1950), p. 37 The restriction to thermal radiation only is common in texts on luminescence, fluorescence, and phosphorescence; also, in E.T. Whittaker, *A History of Theories of Aether and Electricity* (1953), vol. 2, p. 78, where, however, "thermal" and "black-body" seem to be used as synonyms. For the restriction to both thermal radiation and equilibrium conditions see M.W. Zemansky, *Heat and Thermodynamics* (ed. 4, 1957), pp. 101-104. Other restrictions occur in the literature, much as the one imposed by G. Joos, *Theoretical Physics* (London, ed. 3, 1958), p. 620—namely, that the medium is uniform. (For the redundancy of this constraint, see A. Cotton, *Eclairage Electrique*, 14, 405 (1898), and *ibid.*, p. 540, as well as S. Chandrasakhar, *An Introduction to the Study of Stellar Structures*, (Chicago, 1939), pp. 202, 203. Finally, astrophysicists, following Wiedemann, tend to define local (color) temperature as that which fits Kirchhoff's law in equilibrium; see S. Chandrasakhar, *Radiation Transfer* (1950), pp. 7 and 8.

5. G. Joos, *Theoretical Physics' op. cit.*, pp. 620-1. See also E. Mach, *Die Prinzipien der Wärmelehre* (Leipzig, ed. 2, 1900), where Kirchhoff's law seems to be utterly unrestricted at first (p. 140) but then, in view of Wiedemann's refutation, is claimed to be restricted to equilibrium conditions (p. 142).

6. See S. Chandrasakhar, *Radiation Transfer* (New York, 1950), chaps. 10-12, where Kirchhoff's law is shown to be an approximation for gray atmospheres. Contrast this with A.B. Pippard [*The Elements of Classical Thermodynamics* (1961)]:

"The success of thermodynamics in these circumstances [of cavity radiation] is perhaps the strongest evidence we possess for regarding the laws as valid in all physical situations to which they can be applied." Another strange phenomenon is the proof of Kirchhoff's law given in A. Sommerfeld's *Thermodynamics and Statistical Mechanics* (1956), p. 135. There is no follow-up on it that I know of.

7. See J. Agassi, *Towards an Historiography of Science* (1963, 1967).

8. Note that in exceptional cases the tinkering goes too far rather than not far enough, when writers apply the "conventionalist twist" [see K.R. Popper, *The Logic of Scientific Discovery* ()], and turn the law into a tautology. In our example the "conventionalist twist" is given by *Handbuch der Physik* (2), which renders Kirchhoff's law a definition of thermal radiation. However, "in practice, the luminescence emissions and thermal radiations of phosphorus are readily distinguishable because there are large differences between their ... characteristics" [H.W. Leverenz, *An Introduction to Luminescence of Solids* (1950)]. Hence the definition is quite redundant and, with it, the law, as used by *Handbuch der Physik*.

9. I. Newton, *Opticks* (1704).

10. P. Prevost, *Essai sur le calorique rayonant* (Geneva, 1809). For extracts, see D.B. Brace, ed., *The Laws of Radiation and Absorption*, vol. 15 of *Harper's Scientific Memoirs* (New York, 1901).

11. The idea is adumbrated in Newton's *Opticks*.

12. *Phil. Trans.* 92, 365 (1802).

13. *Ann, Physik* 56, 264 (1817). [for English translation see J. S. Ames, ed., *Prismatic and Diffraction Spectra*, vol. 2 of *Harper's Scientific Memoirs* (New York, 1898)]. Note that solar chemistry is impossible without at least some stellar spectroscopy. Hence Fraunhofer's great effort in procuring the spectrum of Sirius and his justified pride in having achieved it.

14. See I. Newton, *Opticks*: "whilst an Assistant, whose Eyes for distinguishing Colours were more critical than mine, did by Right Lines ... note the Confines of the Colours .... And this operation being divers times repeated ... I found that the Observations agreed well enough with one another, and that the Rectilinear Sides ... were ... divided after the manner of a Musical Chord."

15. T. Young, *A Course of Lectures on Natural Philosophy* (London, 1807), vol. 1, pp. 367, 378, 389; see also, *Miscellaneous Works of the Late Th. Young* (London, 1855), vol. 1. Note in the latter work (pp. 82-3) the new theory of harmony of

colors (in Newton's rings) as a new link between acoustics and optics—only to be given up in 1817 (pp. 280-82). See also (p. 35) the beautifully simple relation between ocular biophysics and spectroscopy.

16. Mach [*The Principles of Physical Optics* (London, 1926)] is the only writer, I think, who suggests that Young's optical studies developed from acoustics, through search for an optical analogue to sound interference. But Mach cannot say why the analogy is important. Mach is impressed with Young's study of Newton's rings and with Young's employment of acoustic analogies there, but he fully ignores Young's view of optical harmony in the rings. Also, as a consequence, Mach fails to attribute to Young the idea that, though physically there are infinitely many colors, biologically there are only a few. Peacock, one of Young's two chief biographers, considers the trigger to Young's revolutionary attempt to have been the alleged publication, in 1790, of a nonexistent book by Huygens [see G. Peacock, *Life of Th. Young* (London, 1955)]. Alexander Wood considers the trigger to have been the failure of attempts to detect light pressures [see his *Thomas Young, Natural Philosopher* (New York, 1954)]. This failure was viewed at the time as neither conclusive nor relevant.

17. *Phil. Trans. Roy. Soc. London* Ser. A 92, 378 (1802). The discovery was in some accord with Young's view, but it also demanded modification of that view, as Young admits (*ibid.*, p. 395). Note that here Wollaston and Young see a physical basis for a biological theory of color; but they insist that there are only a few biological colors yet infinitely many physical colors. Young gave Wollaston credit for this discovery, though he himself had predicted it earlier.

18. See C. A. Young, *The Sun* (New York, 1895, 1896, 1898) for plates presenting absorption and emission spectra as negatives and positives; see A. Cotton's impressive "The present state of Kirchhoff's law", *Astrophys. J.* 9, 237 (1899). A list of difficulties in astrospectroscopy is given in *Smithsonian Inst. Ann. Rep. to July 1898* (1900). Concerning the discovery of alleged new elements on the sun, this could not go on indefinitely, in view of Mendeléeff's periodic table, as noted by Sir William Crookes, who, commenting on his own (alleged) new element, considers the game near to a close [*Smithsonian Inst. Ann. Rep. to July 1899* (1901)]. But the problem of identification remained serious for a while. In the strict sense, the problem of identifying solar spectral lines is still not totally solved.

19. *Stockholm Acad. Handl.* (1852-53), p. 229; *Phil. Mag.* 9, 327 (1855).

20. *Trans. Roy. Soc. Edinburgh* 22, 1 (1858). For the priority dispute, see G. R. Kirchhoff, *Phil. Mag.* 25, 258 (1863) and J. W. S. Rayleigh, *ibid.* 1, 98 (1901). See also A. Cotton, *Astrophys. J.* 9, 237 (1899), and H. Roscoe, *Smithsonian Inst. Ann. Rept. to July 1899* (1901), especially p. 621.

21. *Ann. Chim. Phys.* 58, 476 (1860).

22. *Berlin Monatsber.* 1859, 783 (1859); *Poggendorff's Ann.* 109, 275 (1860); *Abhandl. Berlin Akad.* 1861, 63 (1861); *Phil. Mag.* 19, 193 (1860); *ibid.* 21, 260 (1861); D.B. Brace, ed., *The Laws of Radiation and Absorption*, vol. 15 of *Harper's Scientific Memoirs* (New York, 1901).

23. The question of the importance of the hypothesis of the existence of black bodies was seldom raised; see, however, *Sci. Abstr.* 1, 383 (1898).

24. *Wien. Ber.* 79, 391 (1879).

25. *Wien. Ann.* 22, 291 (1884).

26. *Berlin. Sitzber.* (9 Feb. 1893), p. 55.

27. *Ann. Physik* 58, 662 (1896).

28. *Phil. Mag.* 49, 539 (1900).

29. J.W.S. Rayleigh, *Scientific Papers* (Dover, New York, new ed., 1964), vol. 4, p. 483.

30. *Phil. Mag.* 10, 91 (1905); *Nature* 72, 243 (1905); *ibid.*, p. 293.

31. *Proc. Roy. Soc. London* Ser. A 76, 295 (1905); *Ibid.*, p. 545; see H.A. Lorentz, *The Theory of Electrons and its Applications to the Phenomena of Light and Radiant Heat* (Dover, ed. 2, 1952).

32. M. J. Klein, in *The Natural Philosopher* (Blaisdell, New York, 1963).

33. J.W.S. Rayleigh, *Scientific Papers* (Dover, new ed., 1964), vol. 6, p. 45. See also note added in 1911 to vol. 5, p. 253. See also R.J. Strutt, *John William Strutt, 3rd Baron Rayleigh* (Arnold, London, 1924), for this and for Nernst's reply to Rayleigh's letter. In his letter to Nernst, Rayleigh claims priority for having raised a serious problem, and Nernst most agreeably endorses the claim, but carefully prevents the interpretation that Rayleigh's discovery of a problem guided Planck. Even in J.H. Jeans, *Report on Radiation and The Quantum Theory* (London, ed. 2, 1924), where the development begins with the Rayleigh-Jeans formula and the ultraviolet catastrophe and proceeds through Planck's solution to Bohr's theory, it is not quite alleged that Planck solved Rayleigh's problem.

34. M. Planck, *Das Prinzip der Erhaltung der Energie* (Berlin, ed. 5, 1924); *Treatise on Thermodynamics* (Dover, ed. 3, 1945), preface to first ed. and pt. 3.

35. ——*The Origins and Development of Quantum Theory* (New York, 1922).

36. ——*ibid.*, p. 19, and all later philosophical works.

37. This answers the question raised in the opening paragraph of this article. Kirchhoff's law was refuted by Wiedemann [*Wiedemanns Ann.* 37, 180 (1893)], before Planck began his studies, by observations of cold flames (cold from the viewpoint of Kirchhoff's law). This raises the question, Which part of the presuppositions of Kirchhoff should be abandoned? It seems that the question was systematically avoided (see 3). For Planck, however, the situation must have looked very different, since he had a special version of the second law of thermodynamics [see his *Treatise on Thermodynamics* (Dover, ed. 3, 1945), pt. 3.] He could easily accept the proof of Kirchhoff's law, with the added constraint that none of the internal variables were employed in an irreversible process. For, when they are, the second law as he understood it need not be violated together with the violation of Kirchhoff's law. Consequently, the restriction of the law to thermal equilibrium was, at the time, neither necessary nor sufficient; hence Planck's later presentation of the law with this restriction (see 4) resulted from later considerations.

38. O. Lummer and E. Pringsheim, *Verhandl. Deut. Physik Ges. Berlin* 1, 215 (1899); *ibid.*, 2, 174 (1900); *Ann. Physik* 3, 159 (1900).

39. See, for instance 35. In his *Scientific Autobiography* (new ed., 1949), p. 42, Planck speaks ambiguously, saying that in 1900 "it was necessary to introduce a universal constant which I called h." Similarly the comment of Max von Laue (*ibid.*, p. 41) is puzzling. "This finding ... was reported by Max Planck again [!] on December 14, 1900. That was the birthday of Quantum Theory." In a paper submitted for publication in November 1899 [*Ann. Physik* 1, 69 (1900)], Planck estimates $h = 6.885 \cdot 10^{-27} cm^2 \times g \times sec^{-1}$ and mentions importance for a future theory of natural constants. In a paper submitted in March 1900 [*Ann. Physik* 1, 717 (1900)] the estimate is repeated. Planck's estimate on 14 December 1900 is $6.55 \cdot 10^{-27} erg \times sec$. Most latter-day restatements of Planck's findings are encumbered with typographical complications which obscure the fact that h came before quantum theory. It is incredible how much of the abundant transcription and restatement in textbooks and histories is so very mechanical as to preserve even typographical clumsiness. See however, N. J. Klein, *Phys. Today* 19, 23 (1966), especially p. 26.

40. A. Einstein, *Ann. Physik* 17, 132 (1905).

41. D. ter Haar, *Selected Readings in Physics: The Old Quantum Theory* (New York, 1967).

42. For the claim that Einstein's work leads to Wien's law, see E.T. Whittaker, *A History of Theories of Aether and Electricity* (1953), vol. 2, pp. 89, 100-105.

43. See, for example, E.T. Whittaker, *ibid.*, vol. 2, p. 198. A.C.G. Mitchell and M.W. Zemansky, *Resonance Radiation and Excited Atoms* (1934), skip the derivation. The one given in Max Born, *Optik* (Berlin, 1933) is much too cursory. It is customary to add half-life factors to Einstein's formulas [see, for example, *Handbuch der Physik* (2); E. Condon and G.H. Shortley, *Theory of Atomic Spectra* (1951)]. This is hardly permissible [see D. ter Haar (41), pp. 64, 65]. Finally, a piquant item: P. Jordan, *Anschauliche Quantumtheorie* (Berlin, 1936), introduces Einstein's conclusions as empirical results. Even in places where the derivation is correct [for example, in D. ter Haar (41) and M.J. Klein, in The *Natural Philosopher* (1963)], the point of our interest is not explicitly stated: the estimate of the Einstein coefficients is for equilibrium conditions, but, since these coefficients are constants, the estimate is quite universal. The constancy of Einstein's coefficients is, as far as I know, stressed only by Einstein himself [see D. ter Haar (41), p. 171].

44. *Mitt. Physik Ges. Zürich* 18 (1916); *Verhandl. Deut. Physik Ges.* 18, 318 (1916); *Z. Physik* 18, 121 (1917). Einstein's chief concern was the problem of recoil, which is of no interest in the present study.

45. *Phil. Mag.* 8, 293 (1904); *Proc. Roy. Soc. London* 84, 209 (1910). Since Einstein does not refer to Wood but refers only to K. von Mosengeil's study of recoil, it is unlikely that he knew of Wood's work.

46. Einstein himself deduces from Eq. 24 plus Wien's theorem (Eq. 7) both Bohr's $e_m = h\nu$ and $A_{m,n} = \zeta \nu^3 B_{m,n}$, leaving the determination of h and ζ to empirical investigation. This is attractive, since Wien's theorem, as Einstein points out, is purely thermodynamical. Also, this is suspect, since this treatment leaves it as a mere empirical fact that $\dfrac{\zeta}{h} = 8\dfrac{\pi}{c^3}$ All this calls for further investigation.

Einstein's insistence on the fundamental nature of Eq. 7 is Planckian, though Planck would not approve, since Wien's considerations were statistical. For the relations between Planck and Einstein, see M.J. Klein, in *The Natural Philosopher*, vols. 2 and 3 (1963, 1964). See also N. Bohr, H.A. Kramers, J.C. Slater, *Phil. Mag.* 47, 785 (1924); P. Franck, *Rev. Mod. Phys.*, 21, 394 (1949); R.A. Millikan, *ibid.*.

47. A.S. Eddington, *Phil. Mag.* 1, 803 (1925).

Appendix B

The Structure of the Quantum Revolution

Thomas S. Kuhn, *Black-Body Theory and the Quantum Discontinuity.* Oxford University Press, 1978.

B.1 Kuhn on Planck

A century and a half ago James Spedding wrote a two-volume review of Macaulay's Essay on Bacon. That review inaugurated a career dedicated to the study of Bacon, including a biography that is still, perhaps, the best extant, and the preparation of the collected works—by Ellis, Spedding, and Heath—which is still the standard edition in Bacon scholarship. Had I treated Kuhn's new book with a similar degree of attention, the result would be a book-length study of Max Planck and his contribution to twentieth-century thought. Let others take up that task. Here is a mere outline of the task, a few hints. Such a story as Kuhn has attempted to tell is, of necessity, full of loose ends. The choice of which loose end to tie up may be *ad hoc*; the choice is then open to criticism. Or, it may be in accord with some principle; the choice and the principle are then both open to criticism. A detail may be wrongly presented or rightly presented but its role in the story wrongly judged. Nor is this all: the list of what may go wrong and be open to critical comments is already too long. A reviewer must, then, either cover all grounds by a book-length review, or make many questionable choices tacitly. Both options are unsatisfactory, and all that is left for the reviewer is to ask for the reader's sympathy and indulgence and to attempt to be as transparent as possible.

What is the book about? The title suggests the answer: it is the story of how the study of black-body radiation (see below) led to the idea that matter emits and absorbs radiation, i.e., light, not continuously, as required by the wave-theory of light, but in discrete bursts of energy, now known as photons. This idea was revolutionary. Thomas S. Kuhn, author of *The Structure of Scientific Revolutions*, has now written the story of one of the greatest and most exciting scientific revolutions of all time—that of the quantization of light: the quantum revolution.

One might expect this book, then, to be not merely about the quantum revolution but also about its structure. Is it? Most authorities on the matter tend to answer in the negative. This review takes the opposite stand and is thus a tribute to this book.

Let it be noticed that this review is meant to serve as a tribute to this book, since what follows might easily give a misleading impression on account of its being almost entirely critical. Yet there is reason to this: the criticism has to be detailed and elaborate and clearly stated; the praise is obvious in the very attention—critical, as it happens—to all these details, and it invites no elaboration. Moreover, much of the criticism is, in itself, also praise. For example, Kuhn gets carried away by his story, forgets both the evolution and its structure, and gets across a starry-eyed intellectual biography of the revolution's reluctant originator, executor, leader and consolidator—midddle-aged Max Planck. Before entering a critical discussion of Kuhn's story, we may very well notice that his attitude is extremely fair: he offers a new and daring judgement which may, in different ways, both belittle and enhance Planck's stature. He says (p. 171) that Einstein's early researches followed a research program "so nearly independent of Planck's that it would almost certainly have led to the black body law [of Planck] even if Planck had never lived." Kuhn portrays the young Einstein as the only genuine revolutionary amongst physicists, with only one ally for years—young Paul Ehrenfest. Yet Kuhn also reveals, perhaps for the first time, how much the development of the old quantum theory owes to Planck, and to quite a few ideas incorporated in his writings in the period immediately preceding the rise of that theory: Planck was the first to introduce both the quantization of phase-space and transition possibilities. The first idea is essential to the old quantum theory (1913) the second to the new (1925), and both are central in contemporary physics. Therefore, the presentation of the story of the quantum revolution becomes (p. 236) a story about Planck.

B.2   Kuhn on the Quantum Revolution

According to Kuhn, the structure of scientific revolutions is never very clear-cut. Kuhn illustrates this, perhaps, by his story of the many ideas that were proposed during the rise of quantum theory yet never developed; of the many that were proposed and shot-down immediately; of the development of a side-remark tossed by one physicist into a central idea in the studies of another. Yet this is only a part of the story. Each important researcher, Kuhn illustrates fairly clearly, had some ideas which he held more tenaciously than other ideas of his, and these were usually not the ideas put to empirical test. Clearly, quite a few processes evolved simultaneously. Followers of Sir Karl Popper will no doubt be pleased to stress the fact that in the process of intensive research great numbers of hypotheses are tested and refuted and given up. Others, in the wake of Norwood Russell Hanson or of Imre Lakatos or of Gerald Holton, will rather stress the fact that some ideas ran as constant themes through the lists of hypotheses tested and refuted, so that whereas the specific hypotheses were indeed refuted and easily given up, the ideas they variably represented were more stable and not easily given up. Kuhn's position is, in principle, somewhat in between: facts are not so clear-cut.

Nevertheless, Kuhn makes one point in an exceedingly clear-cut manner: one empirical refutation was quite central to the whole story of the quantum revolution. The refutation did not appear in one go, by looking at an experiment and seeing one result when the predictions based on current theory had led to a different expectation. Quite on the contrary, the story of the crucial refutation is complex. It is the story of how it slowly became apparent that the classical, accepted theory imposes on its adherents the Rayleigh-Jeans black-body radiation law which contradicts Planck's, and which conflicts With some very commonly known observations of rather plain facts. Kuhn points out rather forcefully that this was what forced the conservative physicist Planck (as well as H.A. Lorentz) to reluctantly admit the revolutionary character of Planck's law of black-body radiation. The confirmation of that law convinced many researchers that they should take that law seriously, but in order to do fruitful research they had to go further, as a few of them did, and *take the confirmation of Planck's law as a refutation: this refutation of the classical and current theory constituted the very quantum revolution.* Otherwise their research tended to proceed on traditional and thus hopeless lines.

This emphasis on refutations and on the significance of noticing them in order to facilitate the growth of knowledge should warm the hearts of followers of Sir Karl Popper. The emphasis on the continuity of the story, of the fact that researchers were guided by a small number of seminal suggestions which greatly facilitated the growth of knowledge should warm the hearts of followers of Hanson, Holton, and others. Thus Kuhn's story apparently clashes with neither view of science. It only clashes with the view of Michael Polanyi and himself, since in his own view there are no clear-cut refutations in science. It is, of course, to Kuhn's credit that he is carried away and tells the exciting story while attempting to make sense of the details rather than while attempting to force them into his own mound.

In particular, Kuhn illustrates a pattern of thought which has not yet gained the attention of students of scientific method, and it is that researchers repeatedly salvage fragments of ideas of their predecessors, especially of refuted ideas of their predecessors, if and when those fragments fit nicely into their own general and vague preconceptions. This fact seems to fit nicely into Kuhn's theory of the growth of knowledge along fixed paradigms, since a general and vague preconception may very well be what he means to describe as a paradigm. Yet this is not so, since a paradigm governs the community of researchers yet here each researcher has his own pet general and vague idea. If any of those ideas were paradigmatic, it was Niels Bohr's correspondence principle, adumbrated in 1913, stated in 1918, and led to the rise of the new quantum theory in 1925, when new vague and general ideas entered the picture instead. (Even this is an exaggeration, since not all researches of significance conducted between 1913 and 1925 followed Bohr; but there is no need to be finicky here.) Kuhn narrates the story of Bohr's paradigm as that of the aftermath to Planck's researches. Planck's researches themselves, then, as well as those of his contemporaries prior to Bohr's, are, in Kuhn's terminology, the pre-paradigm. Yet a pre-paradigm should be a mere prelude, whereas Kuhn honestly describes the early period of the pre-paradigm as crucial, not the paradigm period.

Perhaps Kuhn is quite consistent here. In his debate with the Popperians (*Criticism and the Growth of Knowledge*) he said, science cannot be all excitement, all revolution: an order must be established before it can be overturned by a revolution. Perhaps this matter has to be left unresolved. After all, the terminology of an established order and a revolution is metaphorical, and to decide matters the metaphor may have to be unpacked first. Nevertheless, if the metaphor helps an author to write a genuinely socially oriented history of science, the metaphor and the author do deserve high praise: it is a rare achievement to write a history of a piece of scientific research as a social phenomenon. Yet, after the homage due to Kuhn is paid, it is time for a criticism of his view—as much too naive.

B.3   Kuhn's Sociology of Science

Kuhn's sociology of science is too little concerned with the genuinely social—with the truly institutionalized. He does notice significant social facts, but they are facts which would normally belong to social psychology, not to sociology proper. And the result is quaint. It has been repeatedly and justly observed that if we consider the social psychology of Erving Goffman (or anyone else's for that matter) as a comprehensive sociology, then we get a highly distorted image of society. In the present case, the picture is clearly meant to be complete and is distinctly odd only because of its claim to be complete. Kuhn repeatedly observes the amazing speed with which developments took place: a theory proposed in England is tested, and its refutation (Popperian-style; Kuhn makes no bones about it) is published in Germany in a few months; A public dispute flares up and the communications get published almost at once.

Intense correspondence and rapid publications support a tight social framework. It is this social framework that makes the following observations of Kuhn both true and significant. researchers with repute lend reputation to each other, to lines of thought, and to research programs. Thus, when Walther Nernst visits Einstein, people learn that Einstein is a person whose views count; when Lorentz endorses the deduction of the Rayleigh-Jeans black-body radiation law from classical theory, the deduction gets the status of rigor; when Planck admits quanta, he legitimizes quanta. Kuhn does not notice the sharp critical spirit of this spirited crowd. But at least he notices what is read and by whom: gone is the old continuity technique which, as I have complained (*Towards An Historiography of Science*), assumes as a matter of course that similarity always indicates influence. Kuhn even notices what standard texts are studied and what proportion of published papers is devoted to black-body radiation, and to related revolutionary topics, such as specific heats of solids and atomic spectra. Assuming exponential growth of the number of researchers, he measures on the exponential scale the rate of growth of numbers of physicists working with the new topics. All this regrettably falls short of sociology proper.

First, as Kuhn himself notices, no matter how reliable a thinker is, a move not worthy of a reliable a thinker is not followed. Second, as Sylvain Bromberger has emphasized, research is not a matter of conviction, but a matter of readiness to invest effort in a project. In deciding on this readiness to join a project, the level of the popularity of the project is a deciding factor for some, but not for others. Also, popularity has an institutional aspect which Kuhn overlooks. Without it, popularity would increase and decrease with no obstacles. Yet, one institution backs another, and this explains both how hard it is to gain popularity for a heresy and how hard it is to discredit an established idea. Thus, when a popular research program becomes established, a researcher may feel obliged to follow it without conviction. This is what happened to Nernst. Kuhn erroneously assumes that Nernst got converted, despite evidence to the contrary; the error seems to be the result of the oversight of the institutional aspect of popularity. This aspect of popularity explains researchers' struggle for institutionalization—for the establishment of the popularity of research projects or research ideas. Boltzmann and Planck spent much effort this way; Einstein did not. Their conduct has to be explained in the light of this fact. In particular, as we shall see, Boltzmann was eager for Planck to legitimize his (Boltzmann's) ideas.

On this aspect, of the search for popularity, Kuhn is enigmatic, fragmentary, or in plain error. In particular, his statistics are misleading, since he does not show that it may just as easily apply to an unsuccessful fashion as to a successful one, so that his theory offers him no tool for the distinction between them. He himself notices that the wide spread of the classical view electrodynamics and statistical thermodynamics—was delayed; he does not say why; nor can he explain the spread (in the second half of the nineteenth century) of those fashions which were later superseded and have left little or no trace in today's up-to-date science textbooks, particularly non-field (Ampére-Weber-Duhem-Ritz style) electrodynamics.

Kuhn is still too much impressed by the up-to-date science textbook, even though he is willing to correct an omission of a significant point of fact here and an oversight of an intriguing line of thought there. His high establishment view leaves him too uncritical, say, of James Jeans' outrageous book, the two editions of which served for a generation and more as a general introduction to the subject. Kuhn is critical of other texts only to the extent that they are taken as a tool for the study of their authors' ideas. But he does not criticize the distortions which they contain. Similarly, he notices that J.J. Thomson's light-particles are not quanta, but says nothing about Thomson's own program—to reduce magnetism to electricity—and its failure (on all fronts). One instance of Kuhn's readiness to bow to fame is very important. H.A. Lorentz, the doyen of physics in the early century, deduced the Rayleigh-Jeans black-body radiation law from the classical views and (consequently, it seems) fully and unhesitatingly endorsed it. This caused great disappointment, and his error was at once refuted by commonly known facts. Fortunately, Kuhn reports this development with all suitable detail and emphasis; unfortunately, when he reports Planck's praise for Lorentz's deduction and for his—Lorentz's!—refutation of the deduced law. Kuhn does not even draw attention to Planck's flagrant violation of the truth, let alone attempt to explain it.

But the most questionable aspect of Kuhn's method is that his confinement of his sociology to social psychology makes it very difficult for him to concede that one person can be responsible, systematic, consistent, yet try his hand simultaneously in competing research programs. This is particularly painful when studying Planck's ideas, since clearly Planck was very reluctant to endorse his own ideas as true—a fact which is observed repeatedly by Kuhn. Nevertheless Kuhn "tries too hard to establish the internal consistency of Planck's position ... unwilling to consider the possibility that Planck himself was not always completely clear about what he was doing", to use Martin Klein's words in the multiple review of Kuhn's present book, "Paradigm Lost", published in *Isis*, 70, 1979 (p. 423). Significantly, but from Kuhn's point-of-view inexplicably, Planck wavered often between different guidelines. Thus, the absence of sociology proper in Kuhn makes even his social psychology questionable, and this forces him to violate the historical record too much. Yet his difficulty is real and should not be overlooked. To illustrate this we may look at Klein who, sharing Kuhn's identification of sociology with social psychology, however implicitly, also shares the same difficulty. Klein notices that Kuhn pictures Planck's research as more systematic than it was, and concludes that Planck was less astute than Kuhn assumes (as indicated in the above quotation from his review). Neither of them can accept the common-sense supposition that Planck was pursuing simultaneously and astutely some logically opposing lines of thought. For, they fuse the tentative suggestion with the institutionalized research program: or want of distinction between social psychology and the properly institutional, they confuse the institutionalized with the popularly endorsed.

B.4 Planck's Program

Planck's scientific autobiography includes his own story of what motivated his research in his early days, meaning practically all the researches he was conducting then, and not only those which specifically led him to his study of radiation and thus to quantum theory. He followed, we are told, the footsteps of Hermann von Helmholtz who had held as a central principle the primariness of the law of conservation of energy, alternatively known as the first law of thermodynamics. Planck attempted to grant the second law the same status. What is primariness? What does primariness entail? It entails the conclusion—also stated by young Einstein, incidentally—that any theory based on nothing but these two laws must be taken as certain and as preferable to any alternative theory that makes some specific assumption—which postulates some model, to use the technical term. In terms of a research program—and Kuhn stresses that Planck did have such a program—Planck attempted to explain the conduct of a system by using a phenomenological theory of it, namely, the two laws of thermodynamics and no specific model. Kuhn observes that Planck was very convincing even though he did not quite stick to his program—that his formula was based on certain simplifying hypotheses (p. 151 and note 19 on p. 293 as well as pp. 162-63), and that he added a statistical hypothesis to it only tentatively and only in desperation. This much and not more than this is true in Kuhn's report on Planck's program and progress. But, first, what is primariness?

Newtonian mechanics in general is a program for constructing models in order to explain phenomena, where the models must conform to some general principles (Newton's three laws). Technically speaking, Newtonian research is the proposal of some laws (laws of force) governing some central forces in order to explain some observed facts. (Central forces follow Newton's three laws: they operate between particles in straight lines and depend on distances alone.) In the late eighteenth century it was proven that central forces adhere to the law of conservation of energy. Yet historians of science, with the exception of Alexandre Koyré and very few others, still declare this law a nineteenth-century discovery. Why?

Central forces abide by the law of conservation of energy: any system governed by central forces alone neither gains nor loses energy when it returns to its starting position. This is not true of an electromagnetic system: a dynamo or an electric motor spends or gains energy with each rotation. This was discovered in the early eighteen-twenties. In the eighteen-sixties James Clerk Maxwell published his electromagnetic field equations which abide by the law of conservation of energy only globally, not locally. It was not possible to invent a Newtonian model for electromagnetism, then, though this was shown only by Einstein in 1905, in his presentation of the (special) theory of relativity. Throughout the nineteenth century attempts were made to construct Newtonian models for electromagnetism, both by Newtonians and by others (like Maxwell).

In the mid-nineteenth century Helmholtz came up with a compromise solution: he proposed to take as unquestionably true whatever both Newtonianism and electromagnetic field theories agreed about. This turned out to be the law of conservation of energy, though it took some time before this was clearly seen in this way.

This, then, was the primariness of the law of conservation of energy: it may be used without wondering about the specific model which—according to Newtonianism—validates it in any specific case. This is how J.J. Thomson explained it in 1885. Helmholtz seems to be the originator of this idea. His disciple Heinrich Hertz even tried to go further and give up all Newtonian models as sheer idle metaphysics which can, and so should, be omitted without loss of empirical content. (He also went further the opposite way; his idea that metaphysics is idle influenced mainly philosophers, and his endorsement of the metaphysics behind electromagnetism influenced mainly physicists.)

This, then, was the idea Planck was emulating, as he says: he attempted to elevate the law of entropy, the second law of thermodynamics, to the same primary status as the law of conservation of energy, the first law of thermodynamics. His program was very austere, and phenomenological (in the physicist's sense, not in the philosopher's sense): describe interacting systems, ascribe to each its energy and its

146   APPENDIX

entropy function, and explain their interaction and behavior with no assumption of any model, i.e., with no assumption of any specific (inner) mechanism. He later recorded the hostility to this program among his peers. Indeed, reviews of his *Treatise on Thermodynamics* around 1900 now seem quite surprisingly disapproving.

B.5   The Status of Entropy

Quite clearly, the second law of thermodynamics was viewed with suspicion because it was not a part of the traditional Newtonian machinery: taking it as primary aroused hostility the way the first law did not, since the first law follows, to repeat Helmholtz's point, from Newtonian mechanics. It was not surprising, then, that an attempt was made to deduce the second law from a Newtonian system as well. This was blocked by James Clerk Maxwell, in two formal arguments which prove that the second law cannot follow from any Newtonian equations (or rather any Hamiltonian equation, which is the same for the purpose of the present discussion). First, he observed, Newtonian equations do not change if we change the sign of the time coordinate (so that Monday comes after Tuesday, Tuesday after Wednesday, etc.), whereas the law of entropy does change (telling us that entropy is now less on Monday than on Tuesday, instead of more). The second argument is known as Maxwell's demon. The demon operates a shutter in a screen between two containers, letting fast molecules rush one way and slow ones the other way, thus heating the hotter container and cooling the colder one, thus violating the second law, yet without violating the first.

Formal logic caught up with Maxwell in this century. A valid inference is one for which there is no model (no possible state of affairs) satisfying its premises but violating its conclusion. Before this idea was made clear David Hilbert applied it intuitively to prove independence: one statement does not follow from another when we can construct an interpretation (a model) violating the one but satisfying the other. Still earlier Maxwell offered an example: his first argument is formal and abstract; put in the formal language of models it says: (a cinematic film of) a thermal process run backwards violates the second but not the first law; hence the second does not follow from the first. The second argument says, had the demon existed, he would be able to violate the second law without violating the first. Hence, though Maxwell's demon does not exist, the second law does not follow from the first. Kuhn misunderstands this (p. 31, line 1) in a passage in which he presents his speculation about Planck's program.
(1)

Kuhn claims that Planck gave up the atomic model and endorsed the continuity model. This conflicts with Planck's own report. A historian studying a researcher's work is permitted, of course, to contradict that researcher's report of his motives—but not to overlook it. Here Kuhn severely violates a simple code. Kuhn also offers three items of evidence to support his claim (p. 30). The first item of evidence is a letter of Planck of 1897. Kuhn says Planck held for years the view it expresses; the context, however, shows that view to have been a passing thought at best. More likely, it was a slightly exaggerated presentation of an argument against the current atomic model, not at all an expression of the continuity view. For, it is always legitimate to present a criticism of a view from the position of an alternative view, even when rejecting that alternative. And an argument presented in a letter need not be strictly legitimate anyway. Kuhn's second item of evidence is not from Planck's pen at all. Nor does it support Kuhn's claim on the supposition that Planck would have endorsed it; indeed, everybody endorses it. It is the statement by Helmholtz to the effect that one may consistently reject all models while using models—indeed while using both the atomic and the continuity model. Of course; one cannot consistently endorse both, but one can consistently use both. Moreover, Planck's program was to avoid using models, but he never said it was inconsistent to use them on occasion—as he did when in a pinch. The third and last item of evidence which Kuhn offers in support of his view that before 1897 Planck had become a continuist for quite some time is a musical analogue: both acoustic theory and electricity theory handle resonance. Since acoustics handles discrete media and electromagnetism handles continuous media, however, the analogy goes both ways and is thus no argument here.

It is hard to explain the fact that such a reputed and cautious historian overlooks Planck's own reasonable story in preference to so poorly supported an alternative. Yet the worst is yet to come: the alternative barely makes sense. Planck's program was, to repeat, to present the second law of thermodynamics as primary, as having the same status as the first law, that is. Yet a question remains concerning the logic of the situation. Now that the second law does not follow from the first, does it perhaps contradict it? In isolation it is hard to say. Under certain assumptions they may well contradict each other. In 1896 Planck's assistant, the famous Ernst Zermelo, argued that in Boltzmann's reading—in his statistical mechanics—the two do contradict each other. Planck says, and Kuhn does not notice, that Zermelo's paper made Boltzmann hostile to Planck until, in 1900, Planck made use of Boltzmann's statistical mechanics. The central item in Planck's report is the date: 1900. Kuhn, to repeat, does not notice Planck's report, and claims that the change was made in 1898. Kuhn expresses regret that Planck did not publish the information about the change for two years. Kuhn mentions some historical facts that do not tally with his claim, which facts he dismisses on the facile reason that both Boltzmann and Planck were confused. Arguments of this kind are all too easy to make, especially since hindsight usually offers more clarity than foresight. Hence, any dismissal of information based on claims that informants were confused calls for some strong evidence. Kuhn offers none at all. What is to be done in such cases? Either the claim of confusion should be left undecided or the historian making the claim can be shown as more confused than his allegedly confused heroes.

Kuhn is, indeed, confused about Boltzmann's statistical mechanics. Indeed, Kuhn is unreliable about every point related to statistical mechanics.

B.6 Planck Versus Boltzmann

There is here a small and simple logical exercise which is extremely difficult to explain because, for some recondite reason, neither physicists nor philosophers will attend to the details and, worse, they will imagine the situation differently than described. But try to think of a container and divide it in your mind's eye—not in reality—into two equal parts. Imagine two particles in the container and ask, what is the likelihood that we find two particles in one part, one particle in one part and one in the other, or both in the other? This is a question of distribution; why it is important will be seen soon. First comes the question, how do physicists decide such questions? How does one, generally, determine a distribution?

The answer congenial to twentieth-century methodologists and to some—not to all—statisticians, is hypothetico-deductive: assume any distribution you like, deduce from your assumption observation-statements and test them. The answer given to this question by Laplace and by other classical probability theorists is the principle of insufficient reason: assume all possible cases to be equally probable. Of course, this assumption will be invalid in cases of bias, but bias has to be apparent before deviation from equiprobability is invoked. Augustus deMorgan proved Laplace's idea inconsistent: we can present differently the same possibilities. We may take the question raised here as an example and arrange the possibilities:

(I) once when considering the two compartments and the two particles as identical, then we have two possibilities: then the two particles in one compartment or in separate compartments;

(II) once when we distinguish the compartments and so distinguish between the case both particles are in one compartment from the case in which both are in the other: then we have three possibilities;

(III) once when we distinguish between the particles but not between the compartments: then we fall back to the case of two possibilities, when the two particles in one compartment or in separate compartments;

(IV) if, finally, we distinguish both compartments and particles, then we have four possibilities: one in the one and one in the other, a switch between the two particles, both in one compartment, and a switch between the two compartments. Nor is it *a priori* clear when we distinguish two items when not: two dollar bills, says Erwin Schrödinger, are identical for one purpose and not for another. DeMorgan's logic is impeccable.

Which of the four options is right? The fourth one above is known as Boltzmann's and as the classical option. The second option mentioned above was invented by Bose in 1924 and is known as the quantum option or the Bose-Einstein one. There is a variant of it, no less important than the original, which will be overlooked here. Almost all writers on statistical mechanics take Boltzmann's option as obvious and look for metaphysical excuses for the choice of Bose's option when, in fact, the preferability of Bose's option over Boltzmann's is empirical: it leads to Planck's radiation law (and to other results that are empirically observed). Doubtless, it was empirical considerations

APPENDIX B

which made Boltzmann choose his distribution. For, in empirical fact a gas in a container is never observed to occupy a mere part of the container, and his assumption makes the option of one particle in one compartment most probable, and when applied to a vast number of particles, yields this observed result as almost certain. The idea that since the two particles are different we must attend to the two options of one being here and the other there and *vice versa* is a *post hoc* excuse. Giving it up led to the *post hoc* excuse that quantum theory recognizes no individuality of particles. All this is poor metaphysics, rooted in Laplace's hostility to hypotheses as arbitrary and in his further appeal to the (inconsistent) principle of insufficient reason as a rule for generating hypotheses as if this principle renders them any less arbitrarily.

Planck had an enormous advantage here: he was averse to all models, to all specific hypotheses concerning the inner mechanisms of the systems he was investigating; and he considered each specific distribution a specific model and so, opposing all models, he also opposed all distributions —Boltzmann's included.

To return to Boltzmann: he assumed that a gas moves quickly towards its most probable state and stays there. Assuming this, and assuming that entropy diminishes with the rise of probability, we can explain the second law of thermodynamics by the use of statistics plus mechanics. The computations are no mean feat, but the idea itself is utterly simple, and its simplicity appealed to Planck. Yet, it was not agreeable to him because it makes the second law not primary but dependent on a statistical mechanics. So he began to criticize it. (2) Why should a system tend to move towards the most probable state? What happens in a fluctuation away from this tendency? Can a system in equilibrium, i.e., in the most probable state, deviate from it for a short while? Does the entropy change during such a fluctuation or does entropy represent the average? If the entropy changes, then the second law is violated, since the entropy law forces the system to stay in equilibrium. If not, what stops Maxwell's demon from stopping a system in fluctuation from returning to equilibrium? How often do fluctuations occur? Clearly, the idea was that a gas moves towards the highest disorder, yet disorder is not the same as the most probable state and entropy is something different from both. For, in one sense a given quantity of air is more ordered when all its molecules are in one part of the container, in another sense they are more jumbled there, yet clearly the former state is unequivocally less probable given the distribution law. Uniformity is improbable, but is the sequence 1,2,1,2,1,2 ... more or less ordered than the uniform sequence of ... . 2,2,2,2 ... ? There is no need to answer this question, since relating probability to order is hopeless anyway, since order is less well-defined than probability. Quite clearly the idea of disorder intrudes the way the idea of insufficient reason does, in order to create the impression that hypotheses are not freely employed. They are.

And so, ideas of order are to no avail. The questions posed above, concerning possible fluctuations, troubled Boltzmann and was pushed aside by him. Kuhn well describes Einstein's research program as the program to put fluctuations in the center of research. Planck was aware of the weakness of Boltzmann's attitude to possible fluctuations. At times, Kuhn shows, Planck used Boltzmann's faulty arguments. Yet, on the whole Planck rejected statistical mechanics as much and as often as he could.

Kuhn notices but does not discuss this. Planck's wish to have the second law as primary, yet his readiness nonetheless to follow ideas not compatible with this wish are both systematically ignored by Kuhn. But the logic of the situation is clear. Thus, when Einstein used statistical mechanics in the direction opposite to Boltzmann, and rather than suppress fluctuations in order to protect the entropy law, he went the other way, Planck admitted defeat. Kuhn connects Einstein's fluctuation theory with quanta—wonderfully, and for the first time, really—yet he ignores the fact that Planck's wording of the second law of thermodynamics, namely, one cannot reduce the temperature of a thermally isolated body in thermal equilibrium in order to raise a weight, conflicts with Einstein's view that this, precisely, is Brownian motion: a (random) thermal fluctuation will lift any body, though the heavier the body the less frequently it will be so lifted.

Normal physicists are still confused on this matter; quite generally they refuse to see ideas of great value overturned and so they tend to reconcile them with their successors. On this, Kuhn's philosophy sides with the normal, common-or-garden physicists. Yet in this volume on the origins of quantum theory he usually sides with the extraordinary sharp-eyed, with the clearly logical physicist, who tends to notice inconsistencies between ideas and to emphasize (rather than smooth over) difficulties. Sometimes Kuhn falters. For example, he notices the inconsistency which Einstein emphasized less, namely between Boltzmann's attempts to suppress fluctuations and Einstein's promotion of them to a central position, yet he fails to notice the inconsistency which Einstein emphasized more, between the law of entropy which forces isolated systems in equilibrium to stay where they are and statistical mechanics which permits such systems to fluctuate. Consequently Kuhn cannot explain the reticence which he observes in the scientific community and the paucity of comments there, regarding the logical nexus between entropy and statistical mechanics. He explains this reticence as rooted in the ignorance of normal scientists; yet, clearly, it is not ignorance but suspicion and fear of making an erroneous decision. Kuhn also explains Boltzmann's own reticence concerning the logical nexus between entropy and statistical mechanics as the absence of a clear-cut position. This is odd and calls for evidence. Kuhn's evidence is this: rather than discuss the entropy function and designate it with the letter S as usual, Boltzmann first discusses the function of negative entropy and designates it with the letter E, which he later changes to H. The change, says Kuhn, is meant to have Boltzmann's function dissociated from entropy. Here the explanation offered by Stephen Brush is better: Boltzmann developed his idea about statistical mechanics to the full before identifying the function E, later renamed as H (or rather its negation), with the entropy function from thermodynamics—as a hypothesis. (4)

There are three distinct ideas which Kuhn's story presents in a jumble. One of them is the already mentioned Maxwell's proof that entropy is not mechanical—since entropy is not time-irreversible—i.e., it does change sign when the time coordinate does. The second idea is the answer to the question, are thermal systems in fact reversible or not? The entropy law permits closed systems in equilibrium to be reversible, forbids open systems in disequilibrium from being reversible, and what it says on closed systems in disequilibrium is still controverted. The third is the so-called Loschmidt paradox. Loschmidt argued thus. Let us consider mechanically any system which begins in an improbable state and ends in the probable state which statistical mechanics considers to be the equilibrium state. Mechanics permits reversing the system, i.e., it permits the exit from any equilibrium state along a path of entry into it. Hence, viewed still mechanically, there are as many paths into a given equilibrium state as out of it. Let us now add statistical considerations. We need a set of possible systems and give each equal probabilities. Consider all possible paths leading to a state and view them as equiprobable. Consider the probability of a state to increase with the increase of the number of paths leading to it. It follows that no state is an equilibrium state, since the number of paths leading into it equals that of those leaving it. This is Loschmidt's paradox. Of course, all it proves is the equidistribution over paths, though perhaps intuitively acceptable, has to be rejected on empirical grounds, since it precludes the observed state of thermal equilibrium. Yet Loschmidt's paradox is very important in that it forces one to see that distributions are hypothetical, and may be assumed over entities other than paths. Now Boltzmann's distribution is in phase-space, i.e., over all positions and velocities; and the position and velocity of a particle determine its path in the absence of interaction—at least between collisions. Hence it seems likely but not proven that Loschmidt's paradox hits Boltzmann's distribution. Historically all this was very important. Kuhn notices that Planck spoke for the first time of distributions of states over energies, and that this was a novelty. He notices also that Loschmidt forced Boltzmann to become clearer about the idea of statistical distribution. Yet he conflates three matters, and erroneously blames Boltzmann for the conflation!

Finally a technical point of confusion for experts. Kuhn usually ascribes to Boltzmann less than he deserves, but not concerning Liouville's theorem. In phase space for N particles in an isolated container with 6N coordinates for positions and velocities, each state of a system occupies one point. The point moves on a path determined by the law of conservation of energy (since the system is isolated); the path is on a curved surface in the designated space, on which all systems with the same energy must be found—isoenergetic curves. Systems with different levels of energy will be found on different isoenergetic curves. We can take a small piece of this curve to represent many systems with exactly the same energy and consider this aggregate of points as a fluid, a fluid moving along, and assess the density of such a fluid while it flows: where does it get denser or rarer? Liouville's Theorem says, the density of a "fluid" in phase-space moving on isoenergetic "surfaces" is constant. To relate this theorem to probability theory one needs some assumptions about distributions. But a distribution on systems over an isoenergetic surface is not identical with the distribution of states over one given isolated system, as required if one wishes to apply phase-space considerations to single isolated systems. Any hypothesis that bridges this gap is labelled as "the" ergodic hypothesis or "the" hypothesis of ergodicity (pp. 55-56). "Clearly," says Kuhn (p. 56), "Boltzmann knew the theorem and glimpsed its relevance", meaning Boltzmann knew Liouville's theorem and not the ergodic hypothesis yet felt he could apply the theorem to statistics nonetheless. He did not.

Kuhn finds him at fault. This kind of history is unacceptable: to say clearly, a thinker had a glimpse of an idea he violated, and to conclude from this important corollaries about that thinker's contemporaries (p. 56) and about an error of that thinker that it "was in some part an essential constituent of his viewpoint" (p. 57) is to hang a great weight on a mere thread revealed only by hindsight.

Thus, usually Kuhn finds Boltzmann not sufficiently committed and not sufficiently clear and not sufficiently decisive—and so at fault—yet on the one central (and still open) issue of ergodicity he finds him having "a glimpse" of what he could not possibly know—and so, again, at fault. Yet this is not in any way an effort to belittle Boltzmann. Rather, it is Kuhn's inability to permit a researcher the mental agility required to move freely to and fro between research programs. Boltzmann did make use of diverse ideas and arguments, to varying degrees of precision and of plausibility—and so did Planck; so do all explorers. (3) Yet when the chips were down Boltzmann repeatedly fell upon his claim that a gas in a vessel tends to be distributed in accord with a certain distribution law which explains a number of facts, including its obedience to the second law of thermodynamics. Not surprisingly, Planck found this objectionable. Indeed, when he capitulated he cut out most of the preliminary considerations and said (p. 125), we simply make hypotheses concerning distributions and test them empirically!

B.7 Planck's Capitulation to Boltzmann

Boltzmann and Planck, Kuhn observes, were mighty opponents in an otherwise empty ring. What exactly the issue was, Kuhn cannot say. He notes with amazement that Planck emulated Boltzmann closely, even in details—such as the omission of high fluctuations from the formulas for fear that they may violate the second law—and in the search for the roots of irreversibility in these formulas. He is particularly disturbed, to repeat, at Planck's failure to make acknowledgement to Boltzmann for the introduction of statistics, and for nearly two if not three long and adventurous years; the consequence, says Kuhn, is that Planck's failure to make a proper acknowledgement to Boltzmann is still not corrected by historians (pp. 77-78). He later (p. 98) explains this failure of Planck as due to a reasonable error: according to Kuhn, Planck was converted to statistical mechanics in two stages: in 1898 he was converted to statistical mechanics as a general idea, and only in 1900 was he converted to the combinatorial idea of computing distributions by counting possible states.

This story should be rejected for many reasons, quite apart from its preventing Kuhn from finding out what exactly was the controversy between Boltzmann and Planck. *First* and foremost, there is here the idea that scientific research rests on conversion, that conversion is important for research; there is also here the idea that such a conversion may be initially vague and quite general and later become more specific. Contrary to this, there are so many examples of researchers who tried their hands in diverse directions, who developed ideas which were quite contrary to their deepest convictions, who developed ideas of their opponents in the spirit of self-criticism, in the spirit of criticism, or both. It is time to stop writing the history of scientific research while using religious metaphors. *Second*, Kuhn's story gives the combinatorial approach enormous weight. Admittedly that approach was originally deemed very important for methodological reasons. Kuhn does not endorse these reasons, and this is well and good. But he offers none instead. Nor can he hide behind the claim that, rightly or wrongly, Planck had accepted that approach. He has no ground for his claim that in 1900 Planck was converted into accepting that approach, and he observes himself the fact that in 1901 Planck questioned it and in 1906 rejected it (pp. 121, 122, 125, 186, n. 16: "Planck's abandonment of all references to a theoretical justification of the choice of equiprobable elements in the case of gases is somewhat puzzling ... ."). Planck's own story is the starting point for a view which is nearer to the truth, which takes account of more facts and is less puzzling, and which calls for no excuse for an alleged omission to acknowledge, especially when the alleged omission is from a person like Planck, who was usually meticulous in his acknowledgements. In order to be able to use it we have to acknowledge a simple logical fact—the very difference was between the phenomenological law which Planck endorsed and the statistical theory which he rejected. He borrowed from Boltzmann in successive moves whatever he could, leaving Boltzmann's statistical model itself to the last. Each of these moves only sharpened the difference between the two thinkers until the last move. Once the statistics of radiation was introduced, in the very last move, quantum theory was born. As Planck tells us, at that moment Boltzmann suddenly became friendly. However reluctant Planck was to make this move, the moment he took the plunge and used statistics proper he thereby legitimized Boltzmann's hope that the phenomenological law and the statistical theory are consistent. And, indeed, this was the point that troubled Boltzmann. Planck was aware of the legitimation in this way, and hence he felt obliged, in the name of honesty, to admit to Einstein in 1905 that he was in error, that the victory of the statistical theory was a refutation of the phenomenological law as he, Planck, understood it.

However important Planck found all this, he was still reluctant to use statistics, and considered his use of it a mere stop-gap. In order to apply it to radiation, it is well known, he had to quantize radiation; and so he took the quantization itself to be a mere stop-gap. In 1909, in his Columbia University lectures, and also in 1915, in the English language edition of these lectures, he says he still thinks Einstein is too revolutionary in his proposal to quantize all radiation, to view radiation in general as consisting of particles rather than of waves. But by then he already took Einstein's idea more seriously: in 1913 (English edition 1914) he discusses the idea of light as consisting of quanta. Had Planck capitulated to Boltzmann in 1898, when Kuhn says he did, he could have come to the conclusion which Boltzmann (1895), Rayleigh (1901) and Jeans (1905) had come to (pp. 146, 148). As it was, Planck's change of mind is rooted in the failure of his program, in his admission that a model for his oscillators is needed—which oscillators were developed in 1898-1900 according to Planck's program (see his *Scientific Autobiography*). Kuhn correctly reports Planck's considerations of the system's properties in order to explain the connection in a body in equilibrium between its temperature and the radiation it gives up, or—which is the same—the interdependence of the distribution of radiation (light) in a closed box—or a closed cavity—on the heat of the box's or cavity's walls. For, Planck's researches were developed on as general lines as possible, yet with an effort to specify some formula. When that formula was reached, both a proof of uniqueness of the formula was required and an explanation of its precise physical meaning. First Planck developed his ideas concerning the behavior of the walls of the cavity. But he could not get the full picture. Finally Planck conceded: perhaps some knowledge was needed of the distribution of the energy of the electric resonators in the wall of the cavity which absorb and emit radiation (p. 99, end of first quotation, and p. 100). Kuhn notices that this is the move towards accepting Boltzmann's view—an admission, in brief, of a need for a (statistical) model. In other words, Kuhn's own exposition of a crucial point agrees with Planck's autobiographical narrative and not with his own general idea. We do not need Kuhn's idea at all, then, since Planck's explains almost all the situation's diverse factors (briefly but) well enough. (Especially clear is Planck's incessant use of Maxwell's equations on the assumption of their utter generality.)

To make this clinch, one needs to go over all the technical details of Kuhn's argument and either respond to them or replace them with details congenial to Planck's story rather than to Kuhn's. This would make one rewrite hundreds of highly technical pages. I prefer to leave this task to others.

B.8   Conclusion

Kuhn's work is a pioneering study in that it allows for important changes of opinion, for important yet mistaken ideas which cannot be overlooked without loss of historical comprehension; that these ideas were superseded is not a sufficient reason for a historian to ignore them. Kuhn also presents the growth of scientific ideas at times as the result of an individual's perseverance in the face of public pressure (Einstein's views on the quantization of light first made it hard for him to get a job and then got him his Nobel prize), and at times as the result of a kind of relay race between a few individuals who stimulate each other. He sets a fairly uniform high standard of presentation and of documentation, yet which can be greatly improved by becoming more transparent, more open about one's own weak spots or lacuñae, more inviting to the reader to participate in the author's Odyssey. In particular, it seems that Kuhn is eager to push his philosophy and methodology to the background, since they are controversial, in order to present us with a fairly uncontroversial volume.

In this he has no success. Consequently, the volume is quite difficult to read—but it is worth the effort. But, above all, it is exceedingly difficult to relate the story to any given methodology: all methodologies are abstract—Kuhn's as well as others'—and every study in depth of an episode in the growth of scientific knowledge invites a careful critical examination of them all. What now transpires repeatedly is that every study in depth makes use of the social dimension of the situation studied and that the sociology of science is therefore better introduced explicitly and critically. We still have to wait for properly critical histories of science in which authors describe clearly what their heroes were thinking they were doing and how they disagree with their heroes in a manner that does honor to both.

B.9   Notes

1. Kuhn has been unjustly criticized for not using up-to-date literature which resolves some difficulties which he finds in the classical texts he studies. One can defend him by observing that he is not alone in making the error here criticized: Maxwell is still not understood by many physicists. They often feel Maxwell's demon is problematic—without saying why—and try to present putative laws to forbid his (the demon's) existence. Now, of course, his existence is already forbidden by the second law of thermodynamics. Those who did not notice this thereby prove their lack of comprehension of the logic of the situation.

This should not be confused with Brillouin's idea. Brillouin tried to go into the details of the impossibility of Maxwell's demon. Assuming that the second law is true and that the demon exists and operates a small shutter, what will prevent his success? Brillouin's answer to this question is read as idealistic and as such it is obviously false. But the question much deserves notice, and as an answer to it Brillouin's idea is quite realist, be it true or false. Yet possibly Einstein's theory of fluctuation offers a sufficient answer to Brillouin's question: it is not easy to separate nicely slow and fast molecules. If so, then one may read Brillouin's discussion as the attempt to construct a machine to imitate the demon, and a proof that goes into the details of the mechanism and shows that the machine must fail. (I owe this note to lengthy discussions with Karl Popper, conducted over twenty years ago.) 2. The very assumption that entropy diminishes with the rise of probability suffices to derive Boltzmann's (logarithmic) law in the light of the fact that the entropy of two systems in the sum of their entropies yet the probability that they are in given sates is a product of the probabilities that each of them is in that state (since logarithms are the functions transforming multiplication into addition, of course). 3. An odd fact which has not been commented on is that Stirling's formula is repeatedly used as a bridge between combinatorial calculations and entropy via Boltzmann's theory of entropy as a function of probability. Stirling's initial approximation works well, and the better approximations do not work as well. Why does this trouble no one, despite the fact that in similar cases numerous objections were raised? 4. This observation was added in this version; previously I did not sufficiently appreciate the fine point, though Boltzmann spent much effort on it, as Brush notices.

Appendix C

Quantum Duality

C.1 Introduction and Abstract

The foundations problems of physics are genuine and interesting, yet in order to tackle them we may need to free physics of certain traditional philosophical confusions, especially concerning the relation between successive theories as approximations: the case of different theories as yielding approximately similar results plays an important role both in philosophy and in physics, yet different thinkers view matters differently and extant confusions about approximation invite clarifications.

The clarification may help understand the quantum paradoxes of the wave-particle duality short of the Einstein-Podolsky-Rosen paradox and its aftermath. The background to the contemporary debate on this paradox will be presented here in the light of some methodological clarifications, hopefully in a helpful manner.

C.2 Dismissing Popular Cynicism about Science

Some views about science which are quite popular in diverse circles, especially among sociologists of science, are, doubtless, not very complimentary. It is hard to see why they are popular. The most cynical view of science is the popular, seemingly sophisticated charge that scientists manufacture and package and sell their data on the open market. This is a refined and genteel way of calling them liars and charlatans. And the wrong way to prove that scientists are charlatans is to argue from the true statement that no factual information which science can offer is ever utterly free of subjective error, as the general sociocultural context exerts some influence even on the purest of scientific activity. Also, one may argue quite wrongly from true premise the two distortions are linked by the very fact that the subjective element in scientific information gains legitimacy from the general sociocultural framework. For, whatever is purely subjective usually gets eliminated by colleagues fairly quickly; it is the generally and forcefully championed error that prevails. The correct view that neither fact nor fiction is free of subjectivity, leads many sociologists of science to argue with all the mock-seriousness they can muster, for the incredible idea that scientific fact is but a special form of fiction: what is not purely objective, they say, is fiction; assume this, and assume that nothing is purely objective, and you may conclude that every allegation is fictitious—that hence science too is fiction.

There is a grain of truth to this fantastic conclusion. No doubt, on the one hand the authors of works of fiction often clothe their narratives in convincing fact-like garbs, such as the form of a novel comprising fake letters written in different styles to reflect different—fictional—letter-writers, such as linking fictitious events to historical ones, and more. Hence, even fiction includes some facts, some non-fiction. On the other hand, no doubt, the personal styles and private prejudices of great researchers like Sir Isaac Newton or Niels Bohr, cannot fail to exercise tremendous influences on scientific research. Hence, no fact is utterly free of some fiction. Yet to conclude that inherently science is the act of convincing people the way artists create their illusions is not serious, and declaring them peddlers in exciting beliefs like itinerant preachers is nothing short of insult.

One fact which may explain the popularity (even among scientists) of this unserious, insulting idea, that science is inherently no different from fiction, is the fact that until recently most empirical research scientists were deeply convinced of the orthodox view that science is fully objective, as it allegedly insured the utter certitude of the most exact, most carefully obtained data. The new folly helps them jettison this orthodoxy. Yet to swing from one exaggeration to another is quite unnecessary and quite aggravating.

Let us notice that at least until recently, probably to date, the tool-kit of research scientists is extremely sophisticated in some respects yet extremely naive in others. The picture of the situation within science gets amazingly quickly sophisticated. To mention but mathematical sophistication, we may notice that the most sophisticated mathematical tools available in one generation are inadequate for the next, and that this is so for well over a century. Yet the unsophisticated part is no less amazing. For example, every research scientist wants to get published, and fast. And she learns the ropes of getting published incidentally to her work because she thinks that as long as she has something which seems to her new and worthwhile she will have no trouble getting published. Yet publishing is not science, the organization of the publication of scientific novelties is not scientific, and many scientists are deeply disappointed to find old and uninteresting materials oft getting precedence over exciting ones in the leading scientific press, that gossip and general impressions are not sufficient for keeping abreast, that one must know more about the quite unscientific setting within which science proceeds. This means that scientists may improve their efficiency by losing their naivete in matters philosophical, psychological and sociological.

The diagnosis of all this is fairly complex, yet the etiology—the root cause—is fairly obvious: understandably excessive optimism, leading to quite naive faith in the utter objectivity of science. First, it was said, any industrious researcher is assured of finding some facts, every field with sufficiently many facts will be assured of founding on them genuinely scientific theories. This is the exaggerated view against which many sociologists and philosophers of science declare scientists to be con-men and mountebanks and charlatans—since no one is assured of finding new scientific facts or new scientific theories or certitude, no matter how hard one tries. We have cases of people who were very successful once, but who tried hard to repeat their success in vein. Still worse, there is the view that a really good researcher must be bold and make up a guess, but the intuition science employs must be correct: guesses made by good scientists must be confirmed.

Is it not excessive to require that one makes guesses and then proves them true? It is; and one who lives up to this excessive requirement is suspect. How else? Moreover, if one is cautious and does not claim to be living up to these requirements, while others—other researchers, friends, historians of science, the scientific establishment—if others say of one that one lives up to these high expectations, then unwittingly they thereby throw doubts on one's integrity.

Consider some historical examples. We are told that Albert Einstein's theory of the distribution of radiation, his theory of the photoelectric effect, so-called, was confirmed by experience, by measuring the photoelectric effect; we are told similar stories about the theory of Max Planck. Indeed, the Nobel Committee which refused to grant Einstein the coveted price for his theories of relativity on account of their speculative nature, granted him the price on account of his studies of the photoelectric effect which proved to be solidly empirical facts. This is why Einstein did not go to Stockholm to receive the price. Now, was Einstein's study confirmed? Could both Einstein and Planck be right? Einstein's considerations led to Wien's law of the distribution of radiation, which contradicts Planck's. Only decades later, when Bose wrote to Einstein about a new distribution hypothesis, the Bose-Einstein statistics so-called, only then could Einstein's early considerations be modified so as to lead to Planck's radiation law. Yet even Planck's radiation law was stuck—until Bohr's model of the hydrogen atom appeared. Even Einstein, who alone declared all light to be discrete, not only some emissions and absorptions, even he could not go beyond certain elementary results until Bohr made discrete the energy levels of the electron in the atom. Bohr's theory, incidentally, was highly confirmed. Yet no one ever took it to be true. On the contrary, it was known *a priori* that at best it held only for the one-electron [or one-meson] system—that is, more or less, for the first column of the periodic table, the ionized second column and the doubly ionized third.

In other words, when a guess is good, then it is partly correct, not necessarily correct to the full: it is an improvement rather than a perfection. And then the researcher is at liberty to attempt and modify this or that aspect of the best theory extant, in order to make it still better—for, even the very best we have is far from unproblematic and is not likely to be utterly error-free.

This is highly common-sense, and is the simply expression of the view that scientific theories are, when successful, approximations or series of approximations, to more successful ones and hopefully to the truth, to God's blueprint of the universe, to use the metaphor dear to Einstein.

The trouble begins when we face the logical truth that any theory which is not absolutely true is false. This is no trouble for the idea of ever improved approximations, since it is the idea that some mistakes are worse than other mistakes.

Einstein said, for example, general relativity can never be taken seriously as more than a mere stop-gap: it assumes both space or fields and matter or particles, and this is excessive. But this same logical truth (that partial truths are falsehoods) is troublesome indeed for those who make excessive claims for science as always in the right. The question is, why do they make such naive and excessive claims? This question will not be discussed here. Suffice it to notice that it looks like a miracle to claim that Yukawa just imagined the existence of the meson and was fully confirmed in his guess; it looks miraculous enough to notice that though as far as we know Yukawa mesons do not exist, something like his mesons do as exist, namely particles with medium mass, with mass between protons and electrons. Yet recently the theory of approximations has been rejected—by Thomas S. Kuhn, one of the most popular philosophers of science, and simply because approximations are not perfect. Indeed, they are not. Kuhn wants both Newtonian mechanics and relativistic mechanics, for example, to be perfect. It fallows that there is no reason for ever giving up the perfection that is Newtonian mechanics. Why, then was it superseded? To this a few answers are given. One answer is that Newton's mechanics is perfect in one sense but not in another. I find this to be a joke in poor taste: it is like telling the perfect lovers that they are, indeed, perfect, but not as lovers. Another answer is that the transition from Newton to Einstein was an irrational act. That makes the foundation of science in principle irrational and scientific schools are thus no more than political or religious sects. To this the same kind of answer is offered: the transition from Newton to Einstein is irrational only in some sense. And so it goes. And this insult to science is very popular today—even among research scientists. The result is great confusion.

    The confusion is on many levels. First, it is the confusion of research on certain lines of approach, of research based on some basic ideas, with the conviction that these basic ideas are true. Examples to the contrary are very well-known. When Michael Faraday denied the very existence of particles and viewed what we usually call particles as points of singularity in the field, this did not stop him from outlining a mathematical theory of dielectricity in terms of dipoles of particles—a theory worked out soon afterwards by young William Thomson, later better known as Lord Kelvin. When Max Born learned of Schrödinger's equation, which he considered a mere mathematical tool but not a contribution to our comprehension of the phenomena, he both explained its importance by interpreting it statistically and used it in a manner very agreeable to Schrödinger himself, by developing his theory of quantum scattering. And all his life he took this as evidenced for his being open-minded. Open-mindedness, then, is developing as best one can even ideas one thinks are false. It is, says Bernard Shaw, playing the devil's advocate as best one can. Thomas Kuhn says, in other wards, scientific research and open-mindedness to not mix.

C.3 Approximationism in Action in Modern Physics.

The confusion due to the rejection of the idea of approximations is detrimental to research, regardless of its being an insults to science: science is so strong that it suffers little from the insults which even active scientific researchers direct at it, but their research is not so strong as to escape unharmed. For, accepting approximationism, we may realize that researchers are at liberty to think up any modifications of any good idea, and check later the question, is this or that modification an improvement? And realizing that opens new vistas. For example, Bohr's model of the hydrogen atom was a terrific improvement, yet, *a priori* not good enough. Why else would the old quantum theory have been replaced with the new? Now Bohr's great step forward was his bold quantization of the electron's energy levels in the atom. Now a corollary of that was eery puzzling at the time—the idea of quantum jumps: an electron in a circular orbit of one radius, say, may switch to the adjacent circular orbit of the next available radius, yet it may not, in this case, ever be in an orbit whose radius is of any magnitude between the two. Strictly speaking, then, an electron in Bohr's atom can only move in an orbit; it may vanish on condition that—simultaneously or almost simultaneously—another electron appears elsewhere. It was soon found that the transition was not timeless but took a certain very brief period—10^{-8} seconds, in fact—so that for the transition duration there was no electron at all, perhaps! Even much later theories, quantum field theories to be precise, were at first understood this way: particles—mesons—kept appearing in alternative spots and the frequency of their appearance (and disappearance) was a function of their rest mass.

    Why should one take this aspect of Bohr's atom as sacrosanct? Why can one not view the electron as moving almost instantaneously, as moving very swiftly from one orbit to another? The answer is, to repeat, that this violates Bohr quantization of the orbit, his famous selection-rules so-called. Yet one could easily take the selection rules as approximate! Indeed, no one in one's senses accepts today the selection rules of the old, semi-classical theory except as approximate to the better, fully quantized ones. Yet there was more to Bohr's selection rules than that: his very famous correspondence principle was his version of the theory of approximation or rather his idea of the way it should be applied here: he said, the small scale is the domain of quanta, the large scale is the domain where classical approximations may obtain. Thus, in Bohr's atom, as the orbit gets bigger and bigger, the transition from one to another looks increasingly continuous, and in the limit the free electron behaves classically.

    Not so: the free electron does not behave classically: the Compton effect of the scatter of light quanta by electrons, the Schrödinger equation for the free electron, the Dirac equation for the free electron, and quantum electrodynamics of the electron-photon interaction, all tell us different stories about the conduct of the free electron, but they all agree: the free electron does not behave classically. This does not matter much, perhaps, since to the degree that classical theories are successful Bohr did explain their success. Later theories simply go beyond that to higher degrees of success.

APPENDIX C

This point is oft confused and clarifying it is rewording. The standard reading—the misreading—of the correspondence principle, says, all macro-phenomena, are classical, all micro-phenomena follow quantum mechanics. This is unclear, as there is no single theory of quantum-mechanics Moreover, there is an error lurking here, illustrated by the famous Schrödinger's cat, the cat whose death—a macro-phenomena, no doubt—is a quantum event, as it is caused by a gun triggered by one photon which quantum mechanically had a given choice to be stopped by a grid but, alas! was not. More simply, almost all colors seen in nature are macro-effects only quantum-mechanically explained, not classically explained. To take a simpler example, take a classical corpusclarian theory proper—classical atomic chemistry. Atoms were declared absolutely stable, and chemists got away with this because in our environment most atoms are almost absolutely stable. To say that therefore we cannot encounter a piece of matter whose atoms are all highly unstable it is to improperly deduce the erroneous denial of the every possibility of nuclear explosives. We need relativistic quantum theory to explain nuclear explosions, and, no less significantly, relativistic quantum theory is expected to explain the success of 19th century atomic chemistry. Why? Because otherwise the old theory is no approximation to the new. The new theory has to explain the success of the old. Otherwise, the instances confirming one theory refute the other and we may be in search of a third theory to combine them both in viewing both as approximate. For example, Thomas Young's researches in 1800 and 1801 created such a stand-off between the wave and the particle theories of light. The question was, then, could a new wave theory, or alternatively a new particle theory, do that? In 1818, when Fresnel replaced Young's, or Huyghns' longitudinal waves with the by now common but then hard to digest transversal waves, this was achieved to some extent. This is often enough the rule: the ideal approximation turns in fact to be less than ideal and so the search continues: approximationism is a demand, not necessarily a fact—the demand for a comprehensive exploration: a new theory should explain (a) the facts the old one explains; (b) the facts which refuse to fit the old one, and hopefully, (c) the facts adduced as crucial experiments between the old and the new. And when crucial experiments go both ways we look for a new theory. This occurs quite often in physics. Once we view approximation as a *desideratum* which is at times achieved to some extent, and the whole picture of science becomes more reasonable.

Looking at it this way we may easily see how Einstein could in 1905 publish one paper on the assumption that light is a wave phenomenon and another that takes it to consist of particles—without being committed to either. He said he considered his view of light as particles by far the more revolutionary of the two, but he did not thereby commit his credence to particles. On the contrary, from the start he strove for a comprehensive view that should cover the wave aspects and the corpuscle aspect of radiation; he also was the first to seek a theory answering this *desideratum* and to this end he tried to develop in 1909 his so-called needle-radiation theory which was an attempt to combine wave and particle characteristics of light by viewing it as beam-like wave fronts. It did not work, and so he got stuck—until Bohr developed a model of the hydrogen atom.

After that Einstein generalized Bohr's theory with his theory of induced emission, so-called, or his A and B coefficients. This theory gained enormous popularity when it was applied to induced emission proper, to optically pumped systems, so-called, to lasers. Prior to that the theory was played down and scarcely mentioned in standard textbooks on quantum mechanics. Why?

It is a strange fact. The theory is beautiful, very easy to comprehend and master in less than one session yet it was not mentioned in the standard textbook. Why? It is also a strange fact that many physicists hotly deny that the theory was at some time or another suppressed. But this is a plain historical fact very easy to ascertain. (I have myself, incidentally, first met that theory in a history book, not in any standard text, since I studied physics before the advent of masers or lasers.)

Perhaps the question leads us entirely off the concern with waves and particles. But perhaps not. The next move in the story should help decide matters. It is the de Broglie-Schrödinger theory, which presents particles as wavicles, that is to say as complex waves whose energy is almost all centered within a small spatial region which moves on a particle-like trajectory—whose group velocity is the particle velocity. The trouble with this theory is very well-known: wavicles may—and often do—dissipate, yet electrons do not; and when we apply this to photons, they always dissipate, much too fast to look like particles. And here, with the coherent waves of induced light we have a perfect case of a de Broglie-Schrödinger corpuscle-like wave. Is this somehow linked to our problem? Why is coherent light—the Gaussian wave packet—able to behave semi-classically? Once we have a general answer to this which fits the phenomena, we have resolved the wave-particle problem in a manner which renders all prior theories approximations—at least as far as photons are concerned.

To return to Einstein's A and B emission and absorption coefficients. B. L. Van der Waerden observes in his *Sources of quantum Mechanics* of 1967, that Bohr's theory had no transition probabilities and that these were the next central target, first attacked by Einstein so that, naturally "all subsequent research or absorption, emission and dispersion of radiation was based upon Einstein's paper" of 1917. The reason the breakthrough was Einstein's, he says, was the founding of its considerations on the idea that radiation is emitted and absorbed in bundles by laws of probability so as to preserve thermodynamic equilibrium (p. 40). In other words, the needle radiation of 1909 is what stood behind Einstein's 1917 original theory of absorption and emission—the idea that a photon has a path. This idea was soon very popular: it is central to Compton's work— his theory and the tests by Bothe and Geiger and by Geiger and Müller, are not comprehensible without the concept of a path of a photon. Einstein's own assessment is expressed succinctly in a letter to Besso (11 Aug. 1916): "Everything completely quantum", and again (6 Sept. 1916): "Every such quantum process is a completely directed event. And as a consequence, the existence of light quanta is as good as assured." The concern here is the exchange of momentum between particle (electron) and field, which assures that field energy is concentrated in small regions—in light quanta. Why should this cause any problem? Looking at various histories, I found no discussion of this point. But the comprehensive *The Historical Development of Quantum Theory* of Jagdish Mehra and Helmut Rechenberg of 1982 offers a clue unawares. At the very end of the second chapter of this book—on the Bohr-Sommerfeld theory of atomic structure, an idea is thrown as an after-thought. In that chapter Einstein's theory of emission and absorption is absorbed into the discussion of Bohr's theory (despite the quotations from letters by Einstein to Besso quoted above), especially into his discussion of his correspondence principle. Yet the chapter ends surprisingly on a dissonance—not between Bohr and Einstein, but rather between Bohr and Adalbert Rubinowicz, who said, "Radiation emitted by an atom is already quantized" (p. 256). Bohr does not so much object to the claim as to the approach: Rubinowicz uses this as a base for the claim—now universally received, by the way—that each photon is polarized from birth. Bohr finds this *ad hoc* and too quick; he hopes to replace it by detailed considerations guided by his correspondence principle. He agrees that both views are of quantum mechanics as a generalization of classical mechanics, but "they approach this goal from opposite sides": Bohr considered Rubinowicz' Einsteinian approach too radical; he hoped to rescue classical radiation theory while sacrificing classical electron theory. "This dualistic approach", we are finally told (p. 257), "certainly contradicted the point of view which Einstein had taken" The chapter concludes by reminding us that Bohr, not Einstein, was the leader at the time. This observation is truly remarkable. Not only is it a solitary sociological remark in that book, it replaces an observation which belongs to physics: it obscures the fact that the justly famous Bohr-Kramers-Slater paper of 1924 refers to the correspondence principle as its guide and explicitly denied photons their paths. But at least as a coda the book cites (p. 358) the fact that in 1922 Bohr was evasive when challenged to comment on Einstein's emission-absorption theory. The question, Pasqual Jordan is quoted to report, pertains not to the application of quantum theory but to its foundations—on which we know nothing. Bohr, thus, clearly disagreed with Einstein on matters of foundations.

Now that Einstein's theory is regularly applied to a broad set of optical phenomena, one can see that there was some justice to Bohr's claim at the time when such applications were missing. And they were missing because of a few errors. One is Kirchhoff's law which conflicts with the very idea of induced emission. The other is one that was not noticed by anyone then—the idea that coherence is different in diffraction and interference experiments, that there is space coherence and time coherence. And coherence is needed in order to have induced emission compatible with thermodynamics! But there is no need to go into that except to say that it fits very well Einstein's idea of a particle-like wave packet—radiation bundle, as he called it in 1917. If this idea, which is speculative and programmatic, can at all have a striking empirical illustration, it may well be the phenomenon of photon echo or photo echo, whose dual name is very telling. Yet Bohr was also right in seeing in a photon path a significant point of principle. To repeat, in 1924 he saw the absence of a path a proposal which can help apply his correspondence principle and develop, quite contrary to the central idea of Einstein's emission and absorption theory, a view of the laws of conservation of energy and of momentum as no more than statistical. This contrast was smoothed over, incidentally, by Bohr's 1924 homage to Einstein's proposal in 1905 to take statistics literally and ontologically. The irony of the situation is in their use of

Einstein's 1905 commitment to statistics: viewing it as encouragement to take the energy and momentum conservation laws as merely statistical, they insulted Einstein by totally disregarding his 1917 claim that strict adherence to these conservation laws is required by statistical thermodynamics. Later on, when the Bohr-Kramers-Slater 1924 theory was refuted by the perfect synchronization of energy loss with energy gain, Bohr was still unwilling to endorse paths. On what grounds? He had no idea between 1924 and 1927. By then he found a reason to deny photons and electrons their paths. This was the path his ideas took from his correspondence principle of 1919 to his principle of complementarity of 1927-28.

Rather than discuss this transition in detail, we may move straight to Bohr's reading of Heisenberg 1927 work as his own principle of complementarity. The idea of wave packets goes very well, of course, with Heisenberg's inequalities, since it helps translate the inequality in Fourier analysis to quantum theory via the Planck-de Broglie formulas correlating the particle and the wave in the ideal, monochromatic case: $E = h\nu$ and $\vec{p} = h\vec{k}$ where $|k| = 1/\lambda$.

Many of the problems involved here can be overlooked, but not the one caused by Heisenberg's reading of his inequalities "uncertainty principle" and his thought experiment known as the Heisenberg microscope; it is now not difficult to dismiss this as a red herring: theories are about facts, not about their measurements. (Heisenberg, too, protested faithfulness to Einstein: he claimed his microscope thought experiment echoes the simultaneity thought experiment of Einstein of 1905; he overlooked Einstein's correction of his error in 1917, in his work on general relativity.)

Bohr seized on Heisenberg's uncertainty, claiming that since exact observations of simultaneous conjugate variables—of position and momentum—of one particle is in principle impossible, the very existence of exact values for these is excluded from science, and as metaphysical. And, of course, no exact values of conjugate variables, no path (the path is determined by position and momentum). And, no path, no problem with wave-particle duality. This last sentence was expanded into a whole literature, including a famous book by the then leading philosopher of science Hans Reichenbach.

C.4   The Confused Roots of Complementarity

The principle of complementarity assumes that the reading of Heisenberg's inequalities as pertaining to measurements is necessary, that in that readings the inequalities can never be either qualified or modified, least of all circumvented, and that we then have met a barrier to science unsurpassable. The response of Karl Popper, in his 1935 pioneering work, on science, is simple: we may take this claim as a challenge and try to perform what the theory says is impossible. After all, this has happened in the history of science. Yet the case at hand was claimed a special status for: unlike other, tentative, claims of science, this is final, we are told. Why? Because of the new unique situation: the wave-particle duality. That is to say, not because we have both a wave reading and a particle reading, but because both readings are quite indispensable. Why? Because neither is approximate to the other. Why? No answer. Yet there is one: In Einstein's 1905 theory strong fields are classical, weak fields are quantum, yet in Bohr's 1913 theory large scale events are classical, small-scale quantum effects. So we can take a weak field and a large scale effect and get Einsteinian quanta behaving semi-classically à la Bohr. This case is known as the two-slit experiment. The experiment can be performed. Does complementarily follow? Not necessarily. The situation has to be carefully examined: approximationism does not say all phenomena of a given domain belong to the old theory if not too accurately described. Rather, it says, all events which fit the old theory sufficiently well do so because it is approximate to the better theory, which may exist or else has to be invented. In other words, it is the new theory which should delineate the domain of success of the old, not the other way around, as done by Bohr. Hence, if we say, the success of classical theories is largely in the large-scale strong-field area, much of the mystery is removed. This is not to say that there is no problem here for quantum theory, but that the whole situation invites reconsideration.

Einstein tried hard to refute the reading of quantum theory as preventing particles from having paths; the theory merely ignores paths, he said, because it is statistical. Hence, he said, the theory does not deny them their very existence. The most crystallized argument on this line is the famous Einstein-Podolsky-Rosen 1935 argument. The argument describes a thought experiment in which two conjugate variables of one particle are measured by two detectors which have left the scene long ago in different directions. Niels Bohr denied that the thought experiment proposed in that argument is at all possible: it is impossible in principle, he said, since the instruments constructed to measure one variable of necessity will be in the way of constructing instruments to measure its conjugate. Why? Bohr did not say. What is the link between this newly declared, 1936 physical constraint on constructing instruments and the older, 1928 intellectual constraint on language which imposes a hopeless dependence on both the partially adequate wave-language and the partially adequate particle-language? No answer.

Bohr's reply to Einstein, Podolsky and Rosen of 1935 appeared almost at once—in 1936. Bohr's debates with Einstein were summed up by Bohr in a lengthy, detailed essay contributed to Paul Arthur Schilpp's *Albert Einstein: Philosopher Scientist* of 1947. By 1950 it was the consensus among physicists that Bohr had won the debate. Einstein had hardly any followers. Yet one of these, David Bohm, kept trying. To cut a long story short, the story led to the performance of a variant of the Einstein-Podolsky-Rosen thought experiment, and today many if not most physicists think the experiment has been performed, and it proves Einstein in error. Yet the experiment, after all, not only had been declared impossible to design, it was deemed also impossible to perform, even were it properly designed, because quanta have no paths!

And so, it seems, it is time to reassess the situation. Too much has been said too affirmatively and vociferously and then silently withdrawn while other things were said, also too affirmatively and vociferously. This way the very dignity of science is seriously threatened.

This is no complaint. Ever since science was the basis of tremendous technologies, military and civil alike, and as long as science is still identified with being always in the right, the livelihood of too many people depends on conduct quite unbecoming to the scientific spirit. We need not disapprove of their excessive self-assertion; suffice it to see that those who are still curious will satisfy their curiosity better if they do not take too seriously, and if they do not make, too many, too strong claims.

I shall conclude with one claim that has never ceased to puzzle me—the claim that John von Neumann had a proof. A proof is of a theorem. I do not know what theorem he proved, nor what his proof is. Reference is regularly made to his book—not, as usual, to page numbers or at least a chapter. Nor are there restatements of the proof anywhere. Loosely, the theorem is this: any laws which might govern the paths of quanta contradict quantum theory taken literally. I say "loosely", because I have heard objections to my wording but these did not help me improve it. The experiments performed in the wake of the Einstein-Podolsky-Rosen argument, incidentally, are between quantum theory and the claim that quanta have law-governed paths. In the light of the theorem cited, the experiment is a crucial experiment between conflicting claims: the claim that is quantum theory (as read by Heisenberg) contrary to the claim that quanta do have law-governed paths. The theorem cited should be labelled von Neumann's Theorem; it is named Bell's Theorem. Bell began with certain inequalities in probability theory, and then, assuming no action-at-a-distance, he argued in the wake of the Einstein-Bohr debate that quantum theory leads to forecasts different from the claim that quanta have law-governed paths (on top of the usual laws), because any attempt to measure conjugate variables of a system, even some time after its parts have separated, is blocked by quantum theory but not by the alternative. This means that the quantum prohibition on measurements (as alleged by Heisenberg) persists through time as long as information stays useful, contrary to the reasonable (causal) action-at-the-vicinity (no-action-at-a-distance) alternative. Thus, if two correlated particles separate as free particles, some measurement on them is blocked, but when at least one of them collides with other systems this curse is lifted. It seems clear enough that too much is too questionable here and experiment is premature at best.

It was Bohm who proved that quantum theory is consistent with the ascription of some laws to paths to quanta. (His theory was nonrelativistic—as Schrödinger's was, and only much later was this limitation proved very significant.) He should have thus repudiated the very existence of any theorem to the contrary. That he hit a nerve is evident from Heisenberg's response: Heisenberg denied not the success of Bohm's exercise but its relevance to the debate. Bohm's law for the paths, said Heisenberg, is not testable and therefore metaphysical.

Loosely speaking, Bohm's claim that there are laws of the paths have been tested when the experiment he had designed was performed, and his claim was found erroneous, we are told. Hence, his claim is not metaphysical, since metaphysical claims, as Heisenberg rightly stressed, are empirically irrefutable. More precisely, the excuse of Heisenberg is *ad hoc*: Von Neumann did not say laws of paths are impossible unless they are metaphysical, he said they are impossible—or else quantum theory is false! Von Neumann claimed that quantum theory is statistical, that repeating a quantum experiment to find a statistical ensemble, we have a statistical spread of each variable, and that the product of the spreads of two conjugate variables in the same repeated experiment exceeds a certain magnitude as specified in Heisenberg's inequalities. In Bell's theorem the discussion is limited to paths governed by local variables, not by potentials, so as to exclude the discussion of Bohm's idea, where the paths are governed by potentials.

The case is not of a proof loosely proven by von Neumann and rigorously proven by Bell. The claim that the quantum constraint on measurement is statistical was a point already proven by Max Born. To this Von Neumann added that when we seek a sub-ensemble to narrow the spread of one variable, we broaden thereby the spread of the conjugate variable of the same sub-ensemble. How did he know? By claiming that the same argument that holds for the ensemble holds equally well for the sub-ensemble. This is correct; yet it is hardly a proof. And it is really neither here nor there, since the simplest reading of all this is to say, first, each quantum has a precise path, but repeating an experiment *ad lib* with as high a precision as desired is impossible for two conjugate variables. Hence, von Neumann contradicts Heisenberg: Heisenberg says, there is no path, Von Neumann says there is no law governing any path or else quantum theory is false. Yet, prior to Bell's theorem, all that was known was that quantum theory does not yield predictions exceeding a given limit of precision specified in Heisenberg's inequalities. This was Einstein's reading of the situation. Einstein never quarrelled with the theory in its statistical reading. He only denied that the theory precludes the very existence of precise paths (or, more generally, conjugate variables) and he asserted that these do exist.

The only reason Heisenberg and Bohr had for denying the very possibility of the precise measurement of paths is their wish to deny a quantum its path. Once the path is acknowledged, as it seems to have been with the performance of the Bohm variant of the Einstein-Podolsky-Rosen thought experiment; to say that the results go with Bohr against Einstein is to say only half the truth of the matter: the very possibility of a result goes with Einstein against Bohr. They were both in error if the current reading of the empirical data is correct: once Bohr's reasons for his denial of a path are out we are at liberty to deny also Einstein's path, or to read the experiments to say that there is no path. The road is then clear for rethinking.

The case of the Yukawa meson offers a good analogy. Yukawa's meson does not exist; but some mesons exist, so the fact that Yukawa was not fully successful offers no consolation to those who had hoped to have only two or three elementary particles. Similarly, perhaps Einstein's path or paths must be viewed non-existent, but the old no-path theory is dead and new semi-classical theories spring up these days in all directions. So now some physicists may wish to see the quantum duality resolved in a new and more carefully thought-out manner. The results of efforts in this direction cannot be guaranteed, and if there will be any results their truth cannot be guaranteed, but we can guarantee that any result will be interesting in the extreme, and that philosophically informed, careful analysis cannot impede progress on these lines.

C.5 Conclusion

The central idea here is simple. Neither a wave theory nor a corpuscle theory need be utterly successful in any given, classically delineated domain. Rather, the new theory should delineate the success of any old theory, including that of wave mechanics and of particle mechanics, such as that success happens to be. The new theory, then, need not conform to either the wave or the particle theory: quantum fields should yield quantum waves which should yield older waves as approximate and, when and to the extent that they have stable group velocities, may be viewed as particle-like to some extent. We may hope that the focus of attention should be here: The question may be discussed in detail within quantum theory—as it is or in modification: under what condition can Maxwell's equations be deduced from which quantum equations? Under which conditions is the Heisenberg matrix mechanics valid? Under which conditions the nonrelativistic wave equation deducible from newer, relativistic ones. And, do all these cases explain the empirical success of the old theories such as they are? Once we have a comprehensive theory which covers all the successes of the older ones, the wave-particle duality will cease to be a problem.[1]

---

[1] Martin C. Gutzwiller, *Chaos in Classical and Quantum Mechanics* (Springer, New York, 1990) should be mentioned here, though I lack the competence to comment on it and though it needs no recommendation from me, or from anyone else—as it is clearly an instant classic. In it the way classical mechanics is a limiting case of quantum mechanics is deemed "a fundamental problem" (p. 3), the poverty of most discussions of the quantum thought-experiments is recognized (p. 175), and the oversight of Einstein's 1917 work on quanta for nearly four decades is noted (p. 209) as most remarkable.

BIBLIOGRAPHIC NOTE

For a detailed list of source materials, the references in the final pages of Appendix A above are adequate but too concise, in the style of *Science* magazine, where that appendix first appeared. The reader is advised to consult some of the following works.

For general background, the two volumes of the classical work by Sir Edmund Whittaker, *A History of the Theories of the Aether and Electricity* (volume 1, 2nd edition, 1951; volume 2, 1953, London, Nelson), are the easiest to consult first. The reader should be aware of Whittaker's unusual attitude to physics: in 1951 he still defended Newtonian mechanics as the last word in physics (see my *Towards an Historiography of Science* for details).

The most accessible source material concerning older theories are *Taylor's Scientific Memoirs* and *Harper's Scientific Memoirs*, as well as the famous *Smithsonian Institution Annual Reports*. The contemporary survey that I found most thoughtful and most useful for the study of the problems that led me to write this book is a fogotten paper on Kirchhoff's law by the astrospectroscopist A. Cotton, 1988, English translation published in *Astrophys. J.*, 9, 1899.

As to secondary source for the nineteenth-century theories of light, electromagnetics, spectroscopy and thermodynamics, I recommend the following. For light, Geoffrey N. Cantor, *Optics After Newton: Theories of Light in Britain and Ireland, 1704-1840*, Manchester, Manchester University Press, 1983. For electromagnetics my own *Faraday as a Natural Philosopher*, Chicago, Chicago University Press, 1971, and William Berkson, *Fields of Force: the Development of a World View from Faraday to Einstein*, Routledges, London, 1974. For spectroscopy, William McGucken, *Nineteenth Century Spectroscopy, 1802-1897*, Baltimore, Johns Hopkins University Press, 1969. For thermodynamics the obvious choice is Stephen G. Brush, *The Kind of Motion Called Heat: A History of thee Kinetic theory of Gases in the 19th century*, 2 volumes, Amsterdam, North Holland, 1976, and his *Statistical Physics and the Atomic Theory of Matter From Boyle and Newton etc.*, Princeton NJ, Princeton University Press, 1983. I hesitate to suggest these books because of some omissions in them which are very strange and seemingly at least quite systematic, hidden well behind the façade of his great scholarship and erudition. Yet his scholarship and erudition do make his works useful for the independent reader.

Finally, the quantum revolution. The fullest bibliography for it is in Thomas S. Kuhn, *Black-Body Theory and the Quantum Discontinuity, 1894-1912*, Oxford, Oxford University Press, 1978, here reviewed in Appendix B above. For the history of quantum theory the standard works are Max Jammer, *The Conceptual Development of*

Quantum Mechanics, New York, McGraw Hill, 1966 and Jagdish Mehra and Helmut Rechenberg, *The Historical Development of Quantum Theory*, New York, Springer-Verlag, 5 volumes, c1982. Yet there is no substitute for Planck's and Einstein's marvelously written original works, at the very least their terrific scientific autobiographies: Max Planck, *Scientific Autobiography and Other Essays*, London, Williams and Norgate, 1950. Paul Arthur Schilpp, *Albert Einstein, Philosopher-Scientist*, Harper Torchbook. More original material is easily accessible in B.L. van der Waerden, *Sources of Quantum Mechanics*, Amsterdam, North Holland, 1967. (See also my review of it, "The Correspondence Principle Revisited", *Science*, 3790, vol. 157, August 18, 1967, 794-6.)

NAME INDEX

Ampére, André-Marie ... 62, 106, 143
Angström, Anders Jonas ... 27, 45-46, 67, 73, 83, 120-121
Archimedes ... 113
Aristotle ... 18-20, 30, 66

Babbage, Charles ... 33-34, 37
Bachelard, Gaston ... iv
Bacon, Sir Francis ... viii, 5-6, 14, 25-27, 32-34, 48, 64, 114, 139
Balmer, Johann Jakob ... 55, 90
Bechler, Zev ... ix, 37, 47
Bell, J.S. ... 166-167
Berkson, William K. ... 169
Besso, Michele ... 162-163
Black, Joseph ... 19-20, 22
Bode, Johann Elert ... 90
Bohm, David ... 165-167
Bohr, Niels ... 11, 21, 55, 62, 75, 78, 88-92, 95, 101-103, 106-108, 114-115, 123, 129, 131, 136, 138, 141-142, 157-158, 160-167
Boltzmann, Ludwig ... 10, 83, 88, 93-99, 102, 108-114, 125-127, 129, 143, 147-155
Born, Max ... 133, 138, 159, 167
Bose, Sir Jagadis Chandra ... 113, 129, 148, 158
Bothe, Walther ... 162
Boyle, Robert ... 4, 9, 11, 39, 50-51, 54, 63
Brace, D.B. ... 134, 136
Brewster, Sir David ... 44, 72
Brillouin, Leon ... 112, 155
Broglie, Louis de ... 88, 162, 164

Bromberger, Sylvain ... 143
Brown, Robert ... 112, 150
Brush, Stephen G. ... vii, 150, 155, 169
Bunge, Mario ... 68
Bunsen, Robert Wilhelm ... 67, 69, 71, 75
Burtt, Edwin A. ... iv

Cantor, Geoffrey ... 56
Cantor, Geoffrey N. ... 169
Carnap, Rudolf ... 51, 75
Carnot, Nicolas Léonard Sadi ... 51, 64-66
Chandrasakhar, Subrahmanyan ... 118, 133
Cohen, I. Bernard ... 47, 56
Compton, Arthur Holly ... 61, 162
Condon, E. ... 138
Copernicus, Nicolaus ... 13-14
Cotton, A. ... 133, 135, 169
Crookes, William ... 135

Dalton, John ... 33, 91
Darwin, Charles ... 51
Davy, Sir Humphry ... 33, 64-66
Demokritos ... 13, 38, 49
deMorgan Augustus ... 148
Descartes, René ... 4, 6, 54, 106
Dirac, P.A.M. ... 160
Doppler, Christian Johann ... 45-46, 97
Duhem, Pierre ... 64, 115, 143
Dunoyer, L. ... 133

Eddington, Sir Arthur Stanley ... 59, 130, 138

Ehrenfest, Paul ... 126, 140
Einstein, Albert ... 10, 19, 21, 24, 26, 39, 43, 48-50, 55-56, 61, 78-79, 85, 88-90, 92-93, 95-96, 98-99, 102-103, 105, 108, 110-115, 125, 129, 131-133, 137-138, 140, 142-145, 148-150, 153-156, 158-159, 161-170
Ellis, Robert Leslie ... 139
Euclid ... 105
Evans, G.C. ... 133

Faraday, Michael ... 9, 27, 39, 45, 63-64, 92, 107, 159, 169
Feyerabend, Paul K. ... 112
Fisch, Menachem ... ix
Fizeau, Armand Hyppolyte Louis ... 58
Flügge, S. ... 133
Forbes, James David ... 69
Foucault, Jean Leonard Léon ... 72, 74-75, 81, 120-121
Fourier, Jean Baptiste Joseph, Baron de ... 26, 40, 58-59, 164
Franck, Philipp ... 138
Franklin, Benjamin ... 39
Fraunhofer, Joseph von ... 15, 31-32, 35, 37-40, 42-44, 50, 58, 60, 67-69, 72, 90, 119
Fresnel, Augustin Jean ... 55-58, 106, 161

Gabor, Dennis ... 112
Galilei, Galileo ... 4, 27, 38, 43, 49, 63, 65, 79, 105
Gauss, Karl Friedrich ... 162
Geiger, Hans ... 162
Gershenson, Daniel E. ... 132
Goethe, Johann Wolfgang ... 55, 59
Goffman, Erving ... 142
Goldstein, Samuel ... ix
Goudsmit, Samuel Abraham ... 32
Greenberg, Daniel A. ... 132
Grimaldi, Francesco Maria ... 53
Gutzwiller, Martin C. ... 168

Hamilton, Sir William Rowan ... 146
Hanson, Norwood Russell ... 140-141
Heath, Douglas Dennon ... 139
Heisenberg, Werner ... 111, 164, 166-168
Helmholtz, Hermann von ... 27, 55, 71, 90, 107-108, 126, 144-147
Heraclitus ... 16
Herschel, Sir John ... 34, 40, 44, 61, 72
Herschel, Sir William ... 22, 34
Hertz, Heinrich ... 98, 145
Hilbert, David ... 133, 146
Holton, Gerlad ... 140-141
Hooke, Robert ... 9, 54
Huygens, Christiaan ... 34, 36, 54, 119, 135, 161

Jammer, Max ... 170
Jeans, Sir James ... 100-102, 126, 136, 141-143, 153
Jenkins, E.A. ... 133
Jones, R.F. ... 104
Joos, Georg ... 117, 133
Jordan, Pasqual ... 138, 163

Kant, Immanuel ... 13
Kelvin, William Thomson, First Baron ... 90, 106, 159
Kennard, E.H ... 133
Kepler, Johann ... 13, 54-55, 89
Kirchhoff, Gustav Robert ... vi-ix, 14-15, 23-25, 42, 44, 49, 66-75, 77-78, 80-92, 94, 97, 102, 108, 110, 113, 117-119, 121, 123, 125, 127, 129, 131-135, 137, 163, 169
Klein, Martin ... vii, ix, 93, 100, 132, 136-138, 144
Koyré, Alexandre ... iv, 46, 69, 103, 145
Kramers, Hendrik Anthony ... 138, 163-164
Kratylos ... 16

NAME INDEX

Kuhn, Thomas S. ... vi-vii, 5, 17, 64-65, 106, 115, 139-143, 151-155, 159, 169

Lagrange, Joseph Louis, Comte ... 25, 63
Lakatos, Imre ... vi, 24, 69, 102, 140
Laplace, Pierre Simon, Marquis de ... 13, 63, 148-149
Laue, Max von ... 137
Lavoisier, Antoine Laurent ... 9, 17, 20, 64-65, 85
Le Sage, Alain René ... 25-26
Leibniz, Gottfried Wilhelm ... 63
Lenard, Philipp ... 99, 110, 114
Leslie, Sir John ... 15, 102
Leverenz, H.W. ... 134
Liouville, Joseph ... 151
Locke, John ... 50
Lockyer, Sir Joseph Norman ... 71
Lorentz, Hendrik Antoon ... 15, 98, 126-127, 136, 141-143
Loschmidt, Josef ... 151
Lummer, O. ... 137

Macaulay, Thomas Babbington, Baron ... 139
Mach, Ernst ... 32, 35, 37, 40, 50-51, 133, 135
Malus, Etienne Louis ... 57-58
Maxwell, James Clerk ... 9-10, 62, 92, 95-96, 101, 104-109, 112-114, 125-127, 145-146, 149-150, 154-155, 168
McGucken, William ... vii, 69, 89, 169
Mehra, Jagdish ... 163, 170
Melville, Thomas ... 34
Mendeléeff, Dmitri Ivanovich ... 135
Meyerson, Emile ... iv
Michelson, Albert Abraham ... 48
Millikan, R.A. ... 138
Mischerlich, Alexander ... 71
Mitchell, A.C.G. ... 138
Mosengeil, K. von ... 138

Müller, Walther ... 162
Musgrave, Alan ... vi

Nernst, Walther ... 136, 142-143
Newton, Sir Isaac ... vii, 8-9, 13, 16-18, 20-22, 25-26, 31-32, 34-43, 45, 50-51, 53-57, 59, 63, 67, 73, 85, 89, 104-108, 113, 115, 118-119, 134-135, 145-146, 157, 159, 169
Nietzsche, Friedrich ... 94
Nobel, Alfred ... v, 96, 98, 103, 114, 132, 154, 158

Odland, Lance ... ix
Oersted, Hans Christian ... 63

Parmenides ... 13, 49
Paschen, Louis Carl Heinrich Friedrich ... 84
Pauli, Wolfgang ... 111
Peacock, Thomas ... 135
Pico della Mirandola, Giovanni ... 105
Pippard, A.B. ... 133
Pirandello, Luigi ... 8
Planck, Max ... vi, 6, 10, 15, 24, 31, 47, 50, 65, 73, 84-86, 88, 90-95, 98-103, 105, 107-114, 117-118, 122, 125-129, 131-133, 136-154, 158, 164, 170
Plato ... 103-106, 110
Podolsky, B. ... 156, 165-167
Poincaré, Henri ... 115
Polanyi, Michael ... 141
Popper, Sir Karl ... iv-v, 5, 27, 33, 45, 65, 70, 85, 87, 91, 102-103, 112-113, 132, 134, 140-142, 155
Poynting, J.H. ... 133
Prevost, Pierre ... 15, 23-28, 39-40, 62, 66, 73, 78, 86, 102, 118, 120-121, 123, 132, 134
Priestley, Joseph ... 20, 62
Pringsheim, E. ... 133, 137
Pythagoras ... 55, 89

Rayleigh, John William Strutt, Baron ... 100-102, 109, 111, 114, 125-126, 135-136, 141-143, 153
Rechenberg, Helmut ... 163, 170
Reichenbach, Hans ... 164
Richtmyer, F.K. ... 133
Ritz, Walther ... 143
Röntgen, Wilhelm Konrad von ... 57
Roscoe, H. ... 135
Rosen, Nathan ... 156, 165-167
Rubinowicz, Adelbert ... 163
Rumford, Benjamin Thompson, Count ... 64-66
Rutherford, Ernest, First Baron ... 88, 90-92
Rydberg, Johannes Robert ... 90

Sabra, Abdelhamid Ibrahim ... 54
Sachs, Mendel ... ix
Schilpp, Paul Arthur ... 133, 165, 170
Schrödinger, Erwin ... 89, 148, 159-162, 166
Segre, Michael ... ix
Shaw, Bernard ... 159
Shimony, Abner ... ix
Shortley, G.H. ... 138
Slater, John Clark ... 138, 163-164
Socrates ... v
Sommerfeld, Arnold ... 134, 163
Spedding, James ... 139
Stahl, Georg Ernst ... 20
Stark, Johann ... 45
Stefan, Joseph ... 83, 95, 97, 99, 102, 125
Stewart, Balfour ... 73, 120-122
Stirling, James ... 155
Stokes, Sir George Gabriel, Bart ... 61-62, 72, 74
Strutt, R.J. ... 136
Swan, Sir Joseph Wilson ... 67
Szilard, Leo ... 112

Tait, Peter Guthrie ... 107

Talbot, William Henry Fox ... 69-71
Taylor, Francis ... 169
ter Haar, D. ... 137-138
Thales ... 12-13, 18
Thomson, Sir J.J. ... 11, 90, 92, 107, 133, 143, 145
Tyndall, John ... 27

Vaidman, Lev ... ix
van der Waerden, B.L. ... 162, 170
von Neumann, John ... 166-167

Weber, Wilhelm ... 143
Wheatstone, Sir Charles ... 45, 67
Whewell, William ... 27, 34, 70, 103
White, E.H. ... 133
Whittaker, Sir Edmund ... vii, 93, 115, 133, 138, 169
Wiedemann, Gustav Heinrich ... 84, 86, 133, 137
Wien, Wilhelm ... 73, 87-88, 93, 95-100, 102, 108-110, 113, 117, 125, 127-129, 138, 158
Williams, L.Pearce ... ix
Wollaston, William Hyde ... 22, 32-40, 42, 50, 55-56, 68, 119
Wood, Alexander ... 135
Wood, Robert William ... 48, 78-79, 129, 138

Young, Thomas ... vii, 9, 33-37, 39-40, 43-44, 46, 49-51, 54-56, 58, 119, 134-135, 161
Yukawa, Hideki ... 159, 167

Zeeman, Pieter ... 45, 98
Zemansky, M.W. ... 133, 138
Zermelo, Ernst ... 147
Zimmermann, Benno ... ix

SUBJECT INDEX

Absorption ... 11, 15, 28, 61, 74, 162
Absorption coefficients ... 74-82
Absorption mechanism ... 87
Absorption spectra ... 8, 29-30, 34, 41, 43, 68
Achromatism ... 39
Acknowledgement ... 39
Amateurs ... 71
Amplitude ... 58-59
Anti-intellectualism ... 32-33
Approximations ... 43, 85-86, 100, 102, 113, 156, 158-162, 165, 168
Astronomy ... 45
Astrophysics ... 43-44, 70, 75, 86, 133
Atomic spectra ... 69, 72, 89, 91
Atomism ... 88, 91-92

Bias ... 33
Biology ... 51
Black body radiation ... 81-84, 86, 88, 94, 96-98, 108-110, 114, 139-143
Bootstrap operation ... 71

Caloric ... 20, 66
Certitude ... 42, 91
Charge ... 91-92
Chemistry ... 161
Cold radiation ... 17, 24-28
Color temperature ... 84
Color vision ... 35, 37-38, 55, 59
Complementarity ... 95, 164-165
Complementary colors ... 52
Compton effect ... 61
Condescension ... vi, 46, 63
Confirmation ... 75, 158-159, 161

Conservation of energy ... 107
Conservation of force ... 107
Conversion of energy ... 62
Criticism ... iv-v, 28, 39, 65, 139-140, 142, 147, 153, 155
Crucial experiment ... 36, 113

Death of the universe ... 104
Degrees of freedom ... 101, 114
Dielectricity ... 159
Diffraction ... 53-55, 57, 60
Discovery ... 34
Discrete spectra ... 30-31, 35, 38, 67, 71
Disorder ... 87
Dispersion ... 52-53, 60
Dispute ... iv, viii, 64-65, 87
Distribution ... 44
Doppler's effect ... 45-46, 97

Einstein's coefficients ... 138
Elasticity ... 57
Electricity ... 64
Electrolysis ... 92
Electromagnetism ... 12, 15, 60, 62-63, 99, 145, 147
Electron ... 11, 89, 91-92, 158-160, 162-164
Emission ... 2, 10-11, 15, 17, 28, 72, 74, 162
Emission coefficients ... 74-82
Emission mechanism ... 87
Emission spectra ... 8, 29-30, 37, 41, 43-45, 68
Energy ... 1, 9-10, 12, 59

Entropy ... 51, 66, 87, 94, 104-106, 108-112, 145-146, 149-150, 155
Equilibrium ... 28
Ergodicity ... 151-152
Error ... viii, 152
Ether ... 9-11, 43, 54-55, 57, 62, 106
Exotericism ... vi
Experience ... 49-51
Experiment ... 5, 32-33, 36, 47-48
Explanation ... 32, 34, 38-39, 41, 49, 51

Fictionalism ... 157
Fields of force ... 107
First law of thermodynamics ... 65, 107, 144-147
Flame ... 16-18
Fluctuations ... 112, 149-150, 152, 155
Fluorescence ... 27-28, 61-62

Gravity ... 62

Harmony ... 36, 54-55, 89
Heat ... 8-11, 63-64, 66
Heisenberg's inequalities ... 164, 166-167
History Of Science ... iv-vii, 4-8, 11, 17, 19-21, 26, 32, 44-45, 47, 49, 51, 53, 55-56, 63-65, 68, 85, 102-103, 111, 142, 152-153, 162, 164

Identity ... 9, 12, 16, 18-19, 21
Incandescence ... 86
Incommensurability ... vii, 46, 64
Index of refraction ... 35-38, 40-41, 53-55, 57-60
Induced emission ... 3, 48, 78, 84, 162-163
Infrared light ... 22, 43, 50, 72
Initial conditions ... 13, 106
Instrumentalism ... 42, 51, 96
Intellectual background ... iv, 8, 144, 154

Intellectual frameworks ... 21-22, 56, 107
Intellectualism ... 42
Intensity ... 14, 30, 35-36, 44
Interference ... 36, 53
Interpretation ... 21
Intuition ... 69

Kirchhoff's law ... 14-15, 23-25, 69-70, 73-74, 77-78, 80, 82-92, 94, 97, 102, 108, 110, 113, 117-118, 121, 123, 127, 129, 131-135, 137, 163, 169

Ladder of axioms ... 5
Laser ... 77, 84
Latent heat ... 22, 24, 26
Light ... 1, 8-11, 21, 36-37, 59, 158, 160-162
Logic ... 146
Longitudinal waves ... 57
Luminescence ... 84, 86, 133-134

Maxwell's demon ... 104, 112, 146, 149, 155
Medicine ... 51
Meson ... 158-160, 167
Metaphysics ... 4, 19, 55, 91, 109, 145, 148-149
Middle Ages ... 3
Models ... 14, 91, 101, 103, 105-109, 111-112, 123, 126-127, 144-147, 149, 153-154, 158, 160-161
Molecular spectra ... 72
Monochromatic light ... 44, 58-59, 164

Natural frequency ... 61, 99
Natural width ... 31
Needle radiation ... 161-162
Nuclear piles ... 77
Numerology ... 90

Objectivity ... 156-157, 163, 166
Ontology ... 19, 22

SUBJECT INDEX

Path ... 162-167
Phase ... 53, 84
Philosophy of science ... 32, 64-65, 156-159, 164, 167
Phlogiston ... 9, 17, 19-20
Phosphorescence ... 62, 84, 86
Photochemistry ... 62, 86
Photoelectricity ... 24, 44, 62, 86, 92, 98-99, 110, 112, 114, 158, 160
Photography ... 43-44
Photon ... 162-164
Photon echo ... 163
Photosynthesis ... 24
Physiological optics ... 33-34, 36-38, 50, 52, 55, 59
Physiology ... 50
Platonism ... 103, 105-106, 110
Polarization ... 57, 60-61
Positivism ... 70
Precision ... 70
Prejudice ... 20, 26-27
Prevost's law of exchange ... 23-28, 118, 120-121, 123, 132, 134
Probability ... 3, 75, 148-149, 151, 155
Public relations of science ... 94, 154

Qualities, primary and secondary ... 49
Quantity of heat ... 63-64
Quantization ... 160, 163
Quantum duality ... 156, 163-164, 167-168
Quantum exclusions ... 114
Quantum jumps ... 92, 160
Quantum theory ... 3, 10, 60, 62-63, 66, 72, 78, 92, 95, 98-101, 105, 108, 110-111, 113-116, 140-141, 149-150, 153, 160-161
Questions ... 3-9, 19, 24-27, 35, 60, 147

Radiant heat ... 22-23, 26, 44, 66
Radiation ... 1, 133-134, 136, 138
Radiation theory ... 1, 3, 7-8, 10-12, 14, 16, 23, 48-49, 98, 101, 139, 141-142, 158

Radioactivity ... 77-78, 91-92
Rationality ... 41
Reflection ... 1-3
Refraction ... 14, 60
Repeatability ... 4
Resonance ... 1, 3, 61, 72, 74, 80, 89, 99, 138, 147
Reversibility ... 80
Royal Society of London ... v, 5

Science ... 16, 47, 157-158
Science textbooks ... 85
Scientific method ... 24, 71, 87, 89, 99
Second law of thermodynamics ... 65, 82, 97, 104-105, 107, 110, 112-113, 117, 122, 124, 126-127, 132, 137, 144-147, 149-150, 152, 155
Semi-classical quantum theory ... 160, 162, 165, 167
Sensations ... 49-51
Simplicity ... 12-13, 54-55, 145, 149
Simplification ... 85
Sociology of science ... 157
Solar spectrum ... 41-43
Sparks ... 45, 67, 71-72
Spectra ... 6, 14, 29-31, 33, 43, 45-46, 52-55
Spectral analysis ... 67
Spectral lines ... 14, 29-35, 37, 40-46, 67-71, 83-84, 88-89
Spectroscopy ... 29-32, 34, 36, 38, 41-47, 68, 87
Statistical mechanics ... 99-102, 109, 111-113, 147-153
Statistics ... 129, 134, 138, 143, 145, 147-149, 151-154, 158-159, 162-165, 167
Stellar spectrum ... 43, 134
Stuffed-shirt professors ... ix, 6-8, 47-48, 94, 116
Substance ... 18-19, 22

Technology ... 5, 41-42, 44

SUBJECT INDEX

Testability ... 70
Thermal equilibrium ... 22-23, 28, 81-83, 86-87, 97, 100, 104, 109, 124, 137, 149-151, 154
Thermal radiation ... 23-25, 28, 86, 117, 133-134
Thermodynamics ... 15, 63, 143-147, 149-150, 152, 155
Third Law of thermodynamics ... 83
Thought experiment ... 65-66, 96, 100, 104, 146, 155, 164-165, 167
Tibetan ritual music ... 36
Transparency ... 2, 80
Transversal waves ... 57
Truth ... 64, 66

Ultraviolet light ... 22, 43, 50
Universalism in science ... 20, 117, 127, 132, 144-145
Usefulness ... 41

Vacuum tubes ... 67
Valency ... 91-92

Waves ... 57
Wavicle ... 162
Wien's law of displacement ... 96-100
Words ... 85

Zeeman effect ... 98

MIX
Papier aus verantwortungsvollen Quellen
Paper from responsible sources
FSC® C105338

If you have any concerns about our products,
you can contact us on
ProductSafety@springernature.com

In case Publisher is established outside the EU,
the EU authorized representative is:
Springer Nature Customer Service Center GmbH
Europaplatz 3, 69115 Heidelberg, Germany

Printed by Libri Plureos GmbH
in Hamburg, Germany